FROM OUT OF THE SMOKIES

In Memory of Ernest Koella Trotter

Ernest was the grandson of my wife Brenda,
and she loved him dearly.
He passed away at the young age of thirteen.
Though I never met him, Ernest and I had a lot in common.
He loved the outdoors—fishing and hiking
around Townsend, Tennessee, in the Smokies, near his
father's golf course, Wild Laurel. I wish I could have
enjoyed some of those adventures with him
in the beautiful mountains of East Tennessee—
just as I hope to do with my new grandsons,
Rex and Roman.

FROM OUT OF THE SMOKIES

Stories of Fly Fishing and Life

CHARLIE TOMBRAS

With an Epilogue by Warren Dockter

The University of Tennessee Press
Knoxville

FIRST EDITION.

(*facing photograph*) Left to Right: Rex and Roman Tombras

Library of Congress Cataloging-in-Publication Data

Names: Tombras, Charlie author

Title: From out of the Smokies : stories of fly fishing and life / Charlie Tombras ; with an epilogue by Warren Dockter.

Description: Knoxville : The University of Tennessee Press, [2025] | Summary: "In Knoxville and here at the University of Tennessee, Charlie Tombras is perhaps best known for his nationally recognized advertising firm and his namesake school of advertising and public relations. But Tombras, in this fascinating and well-written memoir, notes that at heart he is a son of the Smokies who spent his childhood in the mountains and streams of Appalachia. Upon his graduation from UT and prior to assuming a position with the Tombras Group, Charlie served a tour of duty in Vietnam as a platoon commander with the 1st Battalion (Airborne), 8th Cavalry. His service earned him the Bronze Star and Air Medal and a rank of first lieutenant. Additionally, Charlie discovered boat racing, parlayed his love of fly fishing into deep-sea adventure fishing, and began ranching, all while growing the Tombras Group to 300 employees and leading the firm to a top 25 ranking among independent ad agencies in the United States. All told, this manuscript presents a captivating look at a life well-lived from one of Knoxville's most successful people"—Provided by publisher.

Identifiers: LCCN 2025001836 (print) | LCCN 2025001837 (ebook) | ISBN 9781621909743 hardcover | ISBN 9781621909736 paperback | ISBN 9781621909767 adobe pdf | ISBN 9781621909750 kindle edition

Subjects: LCSH: Tombras, Charlie | Fly fishing—Great Smoky Mountains (N.C. and Tenn.)—Anecdotes | Conduct of life | Businesspeople—Tennessee—Biography | LCGFT: Autobiographies

Classification: LCC SH549 .T66 2025 (print) | LCC SH549 (ebook) | DDC 799.12/4—dc23/eng/20250610

LC record available at https://lccn.loc.gov/2025001836
LC ebook record available at https://lccn.loc.gov/2025001837

To my new grandsons,
Rex and Roman Tombras

Contents

PART 3: STORM CLOUDS RISING

PART 4: HOMECOMING

PART 5: TRANSFORMATION

PART 6: MEMORABLE CASTS

PART 7: PUSHING BOUNDARIES

PART 8: RIPPLES IN TIME

Illustrations

PHOTOGRAPHS

Following page 141

DRAWINGS

Preface

From Out of the Smokies is a collection of interesting stories from my life, told through the lens of one who fly fished. I grew up on the rivers and trails of the Great Smokies in the 1940s and 1950s, and the lessons I learned in the backcountry and the memories of it became a part of me and who I am. The way those experiences formed me and the dreams that began to grow within me at that early age led to a most interesting life.

The stories in this book trace my growth and what I believed in—as a kid learning to fly fish in the clear, cold headwater streams of the Smoky Mountains, as a teenager racing fast boats on the National Outboard Association circuit, as an airborne ranger, rifle platoon leader in the Vietnam War, and as a whitewater paddler experiencing the thrill of a first descent down a wilderness river in northern Labrador. There are also stories from childhood, school, marriage and family—and the challenges that came my way.

Beyond that and perhaps more to the point, *From Out of the Smokies* is also a seventy-five-year journey of fly fishing shaped by my younger years in the Great Smoky Mountains National Park—a journey that eventually led me out of the Smokies to cast flies in some of the most remote and beautiful places on earth, even on the world's great oceans in pursuit of world-record marlin. In recounting these stories, I avoided the temptation to become too technical, so even those who have never fly fished might enjoy the adventure.

As it turns out, these adventures span my lifetime. Unwittingly, I have written something akin to an autobiography, but that was never the point. I simply wanted to share my stories and perhaps figure out for myself just how I got hooked on fly fishing.

I would like to make it manifest how blessed I was to be able to afford to travel to some of the fly fishing locations in this book. The dream of catching big fish on a fly rod was one of my motivations for working hard and growing our company, the Tombras Group. Otherwise, some of the stories in this book would not have been possible. I am also grateful to my parents, who sacrificed and made sure I got a good education. I am especially grateful to my dad, who taught me the ins and outs of the advertising agency business and for sharing his fly fishing days with me. It was an advantage that not many young boys have, and I am so grateful to him for that.

One final note: The epilogue of this book, which covers the history and growth of the Tombras Group, was written by my good friend and CEO of the East Tennessee Historical Society, Warren Dockter. Aside from the epilogue, all the words in this book are mine. I was tempted to use professional writers but did not. I wanted the reader to gain a more authentic sense of my life, and I felt that no one should tell these stories but me.

Acknowledgments

Several people helped make this book possible. First and foremost, my wife Brenda. We spent days together editing it at her son's (John Trotter's) mountain home with a view of the Smokies beyond—the place where so many of my early adventures began.

The stories in this book are true, and many of them, particularly the fly fishing stories, have been told and retold many times around a fireplace in a 130-year-old ranch house high in the Rocky Mountains of northern Colorado. Good friends gathered there—people like John Emert, Joe Congleton, Edgar Faust, and Graham Hunter. They encouraged me to write these stories.

As I was finishing the book, my good friend, Warren Dockter, PhD, president and CEO of the East Tennessee Historical Society, provided consultation, feedback, and advice on the organization of the material as well as the process of finding a publisher. He also wrote the epilogue, "The Story of the Tombras Group." For all his help and support, I am so grateful.

A special thanks to Brian Potter, creative director at the Tombras Group, for supervising the artwork at the beginning of each chapter and for his encouragement.

Finally, I sincerely thank two medical doctors who kept me alive when the odds were against it: Dr. David Aljadir, University of Tennessee Medical Center, and Dr. Christina Cho, Memorial Sloan Kettering. Without them, this book would not have been completed. Most of this book was written from a hospital bed at Memorial Sloan Kettering, and it was there that my wife Brenda gave me daily encouragement to write. I am so grateful to her for that and for being by my side always.

Introduction

My sister Nancy and I grew up in Knoxville, Tennessee, in a place called Old Westmoreland, on Stone Mill Road. There was a path through the woods to my grandparents' home. It was well worn with use.

My grandfather was a man with an athletic build and the look of one who had done something more than sit behind a desk. His name was Sylvanus Timothy Weaver. Everyone called him *S. T.*

On the weekends he was usually at his farm, between Knoxville and Maynardville, Tennessee, near Norris Lake. The farm was one of those things in life that gave him joy.

A small dirt road rambled through the farm—Weaver Road. From my early childhood, I recall the big white barn and the aroma of musty hay, the cattle and the meandering stream—especially the old mill pond where we walked across the top of the narrow concrete dam early one morning, and I fell off into the deep water. It was before I could even swim, and I was well beneath the surface of the clear water. Instinctively I held my breath, and when I opened my eyes, it was surreal. I had never seen the river from below the surface, and it was startling. For an instant, I was fascinated—and then had the thought that I would drown. But I was not afraid, because my grandfather was there, and I knew he would save me. Sure enough, I felt his strong hand on the back of my suspenders, and he pulled me out.

It was early March, and the water was cold. I was shivering, and he took me to the farmhouse. My grandmother put me in front of the wood-burning stove and dressed me in dry clothes. I felt her love; I called her "Bon Bon." Her maiden name had been Lucy Savage, but now, of course, she was a Weaver.

The old farmhouse was built of white clapboard, and the kitchen opened to a screen porch where my grandmother served Sunday dinners. In the evenings, my grandad lit the kerosene lamps, and I listened to the adults as they talked long into the night. I remember one of those evenings when four of us were there—my great-grandfather, my grandfather, my dad, and me. How often do four generations of men in a family get to spend time together? It seemed only natural, and the thought of it being something special never occurred to me.

My great-grandfather was W. J. Savage. I remember the kerosene lamplight flickering on his weathered face as he placed a tanned case on the table. From it, he removed the sections of something lovely and finely crafted: my dad told me that it was a fly rod and handcrafted of split cane. Carefully and gingerly my great-grandfather laid each section of the rod on the table before us, and we admired it in the lamplight. Then he gave the fly rod to my dad as a gift.

Looking back on that evening so long ago, it's now clear to me who taught my dad to fly fish—and I wonder how far back in time fly fishing had been a part of our family? But I was far too young to ask and now will never know.

Later while I was still young, I held the fly rod in my hand and waved it in the places where the water was pure and clear and tumbled through a forest that was filled with mystery. To me, the rod was like a magic wand, and I cast it over pools and riffles where there was mystery and a hint of hope—and a sense of things yet to come.

The fly rod became a part of me, and it wasn't until years later that I put the rod away for a while. It was a time that belonged to someone else, and I departed from the river to a place called Vietnam. But the river, the fly rod and I were inseparable—and we always reunited.

Our family was like that, too. We loved reunions. Every year on the Fourth of July, there was a big family gathering at the farm in the meadow by the stream, with a picnic, fried chicken, marshmallow roasts, watermelon, music, and fireworks. At least forty or fifty relatives from the Weaver side of the family were always there, and some played banjos and fiddles. It was something I always looked forward to.

I was still a very small boy at one of those family reunions when my great-grandfather showed me a very old motion-picture camera. It may have been one of the first motion-picture cameras in Knoxville, perhaps even the very first. I have film that he had shot at the farm in the 1920s. It's a short film, just a snippet of family, but I treasure it and wish there was more. That film was one of my motivations for this book. I want to give future generations more than just a snippet of what life was like for me when I lived. So I wrote the stories from my life, of the dreams and passions I pursued and the storms that came my way.

To future generations, I hope you and your children and theirs too will have the freedom to live your own stories and that they will be good ones. Your stories and the lives you live will be far different than mine, in ways I can't even imagine.

PART ONE

The Child

THE FIRST THING I REMEMBER, 1945

I WAS THREE YEARS OLD, almost four. It was August of 1945, and my world was about to change. My entire life had been the tiny, red brick apartment on Laurel Avenue near the University of Tennessee campus in Knoxville.

Dad was working for the Knoxville Utilities Board. He had worked there for several years and had finally saved enough to buy us a very small house and a car.

It was moving day from the tiny apartment to the new house that I had not seen and could not imagine. Relatives were there to assist with the move, including my grandparents on my mom's side of the family. They lived just a few blocks away on Highland Avenue. From the Greek side of the family, my aunt Natalie and her sister, Helen, from Chattanooga came to assist. There were many other relatives. They had all come to help with the move to the new house I had yet to see.

I watched the activity with fascination. The big moving van arrived and was parked along the sidewalk. Everyone was helping move clothes and lamps and furniture to the truck. In the air I sensed an energy, a vitality and joy.

My whole world had been this small apartment. It was all that I knew. I had no idea where I was going because I had no frame of reference, no image of any other place. For me, it would be a trip into the unknown. I might as well have been moving to outer space. It was the biggest thing that had ever happened in my three and a half years on this planet and the most exciting day of my life to that point in time. I had faith in my parents that the place we were going would be wonderful.

The scene was teeming with excitement and bounce. All the adults were upbeat and happy. I was standing on the sidewalk near my mom when it happened, and my world crashed: I heard my mom say to my aunt Natalie, "This is such a special day. How unfortunate that little Charlie will never be able to remember it." As I heard those words, I felt a sudden and overwhelming surge of anxiety, and I was distraught.

Mom explained to Aunt Natalie, "I took a class from a professor who helped conduct a major study proving that adults can't remember anything that happened to them prior to the age of four."

This news was extremely alarming to me since I was not quite four years old. How could it be possible for me to forget this day? It was simply the biggest and best day of my life. What a sharp and crushing blow to learn I wasn't even going to remember it. I was devasted.

And then a bit of defiance crept in. "I'll show them."

I decided right then and there that this moving day was so important and such a happy time for our family and me that no matter what, I would always remember it. But how? I was fearful that Mom and her professor might be correct.

I didn't even know what a "study" was, but I came up with a plan to prove it wrong.

It was a simple plan, but pretty good for a little kid not yet four years old. Instead of trying to remember something years later, I would only have to remember it at least once a year. I thought to myself, "Surely I can do that." And I figured if I could just remember it once a year, every year, I would therefore remember it forever.

So, I stood there and watched the scene unfold before me. I strove to remember every visual detail. I wanted to never forget. I took it all in—the strong visual image of my dad rolling out clothes racks and furniture and all the relatives helping him load the moving van. The faded red bricks on the front of the old apartment. The cracks in the sidewalk. The moving van parked on the street with the cab facing west. The sounds of the adult voices, lively and upbeat.

Even after all of that, it still worried me that Mom and her professor might be correct. For weeks afterward, I was fearful that no matter how hard I tried not to forget, I would. The thought of forgetting something that important and exciting in my life distressed me. Perhaps it was the first time in my life that I grappled with the concept that something might not be possible. I needed a plan.

My plan came together in September of that year, on my birthday. Mom reminded me that I was entitled to blow out the candles on a cake and make a wish. This was the breakthrough I needed. Remembering moving day to our new house, and remembering it once a year, every year, was going to be easier than I thought—a piece of cake, in fact.

And every year since, as I blew out the candles, my wish was simple. It was never to forget that moving day. It worked, and even as a child it made me so happy and proud that I had proved the research study wrong.

I drove by the old apartment on Laurel Avenue recently to see if it was still there. To my amazement seventy-eight years later, not only was the old apartment building still there—it was exactly as I remembered.

I stopped my car where the moving van had been. I walked to the spot on the sidewalk where my mom had stood when she told my aunt Natalie that I would never remember this day. For a moment I was that child again and I relived the memory—the faded red bricks of the old apartment, the crumbling sidewalk, my mom and dad with relatives all there. The scene from that wonderful day in 1945 is as vivid in my mind now as I saw it then. And to this day, it is the first thing I remember.

FIRST DAY OF SCHOOL, 1946

I WAS FIVE YEARS OLD, and it was my first day of school. Dad was going to take me there.

"Be sure to go in with him, and help him find his teacher," Mom said.

So I was surprised when Dad stopped the car and let me out at the corner of Kingston Pike and Scenic Drive and wished me well on my first day of school. It was obvious to me right away that there was no school at the corner of Kingston Pike and Scenic Drive.

Even at the age of five, I wanted to appear brave to my dad. But I was shaken and afraid that I might not find the school.

It was called Sequoyah School, and I had never seen it. I couldn't even pronounce the word. I had no idea what it looked like or where it was because it wasn't near our home.

"Where's the school?" I said.

He pointed down Scenic Drive in the direction of Sequoyah Hills and said, "It's that way."

I said, "But aren't you going to take me to the school?"

Dad said, "I'm running late, and it will be good for you to walk to school as I did when I was a boy growing up in Chattanooga."

I asked him, "How will I know what the school looks like?" He said, "Easy. Just look for a lot of kids."

I said, "What if I get lost?"

Dad said, "Just bear left where the road splits."

Unfortunately for me, I didn't know my left from my right. After some discussion and verification on the true meaning of "left," I exited the car and started walking in the direction he had pointed. I kept my left hand entrenched deeply in my pocket to be sure to remember which way was "left."

In about a half mile, sure enough, the road split, and I went left. It felt good to finally get my left hand out of my pocket. But I wondered if I would find the school. And what if I didn't find it? Would I knock on someone's door and ask for help? Would I somehow signal a passing motorist? What then? What would be the outcome? I conjured up all kinds of scenarios and none of them were good.

Still no school.

Not long after that, I saw a red brick building with lots of cars parked in front and to the side—and what appeared to be parents taking their children inside. This must be the school. It had to be.

I walked hesitantly to the entrance. It looked ominous. I remember feeling so different from the other kids since my parents weren't there. This was not a moment of confidence. It was sheer apprehension. What do I do? Where do I go? I felt like maybe I should have asked my dad a few more questions.

Timidly I walked in, avoiding eye contact. I had no idea where to go or how to act. The place was bustling with activity, and every child was being led somewhere by a parent, or a teacher and a parent. I didn't know what to do. I leaned against the wall in the hallway and tried to act like I knew what I was doing. Part of me wanted to be like the other kids and fit in, and the other part of me wanted to disappear.

Finally, a bell rang. The halls cleared, and there I was, by myself in an empty hallway. I don't think I've ever felt more alone in my whole life than I did at that moment.

Eventually a teacher came by and asked me what grade I was in or if I knew the name of my teacher. I didn't know. She seemed surprised at how little I knew and took me by the hand and ushered me into a classroom. By this time the students were already in their seats, and all eyes were on me since I was the only one late.

I felt very small.

Later that afternoon, my mom picked me up in front of the school. When I clambered up into the front seat of our old Plymouth, she was so excited to hear about my first day at school, and she had so many questions.

"How did it go?" "OK."

"What did you do?" "We went to lunch."

"What else did you do?" "We practiced fire drill."

"What else?" "We went to recess. We played."

"Did you study your ABCs?" "What's that?"

"Did you add?" "No."

"Did you subtract?" "No."

She looked at me perplexed and said, "Did you learn anything at all today?"

I thought about it for a minute, and then, excitedly, I said, "Yes. I learned which way is left." And I proudly gestured to the left with my left hand.

She seemed surprised and asked, "Did they teach you which way is right?" I said, "No they did not."

She looked confused and almost disappointed—and we drove home.

The thought occurred to me to tell mom the story of my scary walk to school and how alone I had felt when I finally got there. It was tempting. I so wanted her to know what I had accomplished. But for some reason I could not identify, I decided not to. Maybe it was something just between my dad and me.

Now that I am older, I have wondered if, after Dad let me off on Kingston Pike to walk to school, he didn't follow me in his car and make sure I was okay, just like he watched over me so many other times in my life.

I can't tell you why, but Dad and I never discussed this. I think he was proud of me finding my way to school on my own and probably happy I didn't report it to my mom.

I never brought it up to him because I didn't want him to know how terrified I had been. I wanted it to appear as if it were no big deal and that I was brave, just like he was, when he was a kid in Chattanooga, and he had walked to school. Of course, it was the opposite. I had been scared stiff.

In retrospect, I think it was his way of letting me discover that I could do things for myself. A beginning, however small, of me learning to be independent.

If he were still here, I would thank him.

GREAT SMOKY MOUNTAINS NATIONAL PARK
UNITED STATES DEPARTMENT OF THE INTERIOR
NATIONAL PARK SERVICE
N
W
E
S
GATLINBURG
To MARYVILLE & KNOXVILLE
TOWNSEND
THE SINKS
LITTLE RIVER
WONDERLAND HOTEL
MT. LECONTE
ELKMONT
CADES COVE
SPENCE FIELD
TO ATLANTA
FONTANA LAKE

ELKMONT, 1948

ACCORDING TO LEGEND, the mountains were formed when the giant bird became tired of flying. Searching for a place to land, he sailed lower and lower until his wings pounded the earth forming the valleys and ridges of the Great Smokies. The rains came and the mountains were covered in a mantle of green from which a great forest grew and clear rivers flowed.[1]

The Cherokee were among the first to see the sun rise over these mist-covered mountains. They called it "the land of blue smoke." To them, this land was sacred, a place of peace and tranquility; and in the shadows of the mountains, they felt a sense of permanence. But then another type of human arrived: settlers from the east, just a few at first. But before long there were more and more spilling through the gaps in these mountains and pushing through the narrow river gorges.

Here in the secluded valleys of the Smokies, the settlers cleared small farms, living independent, self-reliant, and private lives. There was a world outside of here, but it didn't seem to be a part of their world. Roads connecting them to that outside world were few, with travel limited by the mountains themselves—and life went on unchanged.[2]

---

1. Inspiration from *Myths of the Cherokee* by James Mooney, 1902, Abstract from the Nineteenth Annual Report of the Bureau of American Ethnology, Washington Government Press.

2. Two preceding paragraphs adapted from a 1966 joint presentation by the author and the writer H. Jackson Brown to the Knoxville Chamber of Commerce to demonstrate the history, heritage, and beauty of the Smoky Mountain area and Knoxville as a wonderful place to live, work, and own a company.

My grandmother Weaver once told me that the roads to the mountains from Knoxville were almost nonexistent when she grew up. Travel from Knoxville to the Smoky Mountain towns of Gatlinburg and Townsend was by train. But change would come, as it always does, and in the twenties and thirties, a new national park was in the works with lands being slowly acquired. Roads would be a necessity.

The Great Smoky Mountains National Park was chartered in 1934, with land acquisitions continuing until the park was officially dedicated by President Franklin D. Roosevelt on September 2, 1940, at Newfound Gap. I was born a year later, so in a way, the park and I are almost the same age, even though the park was officially pieced together prior to that.

Like so many in East Tennessee in my generation, I grew up with the park. Some of my childhood friends had parents who owned cabins at Elkmont and Townsend—the forests and streams were their playground.

My first recollection of fly fishing was in the park with my dad. It was 1948, and I was six years old—too young to fly fish, but I didn't know it. I was excited to be there with Dad. It was a place on the Little River called Elkmont, the evening before the opening day of trout season.

Back then, fishing in the park was not year-round like it is now. Park waters closed in September and reopened the first week in May. The trout had a respite in the winter from anglers, not that there were that many fly fishers back then, and by spring, the fish were very happy.

During this visit to the park, we were staying at a small cabin that belonged to an older gentleman, Bill Newman. His cabin was one of the last four or five on the right as you headed up the gravel road along Jake's Creek. The back of the little cabin had a screened-in porch with an antiquated gas cookstove, a hand-hewn table for meals, a bed with quilts, some cooking utensils, and a rickety old floor. The porch opened through some doors to a small living room with mountain-made rocking chairs and a weathered table battered with time with old kerosene lamps placed upon it. Jake's Creek flowed nearly right under the back porch, and the sound of little one-foot waterfalls and plunge pools gave a hint of adventures yet to come.

My dad and Bill cooked dinner on the porch, and by the dim light of the kerosene lamps, I listened to the stories of their past fishing times together in the Smokies. I didn't understand most of it, but I was fascinated by their talk as Bill puffed on his old pipe and they examined and shared some of their favorite trout flies.

I recall that Bill didn't fish on that trip, yet he was enjoying those few days in Elkmont just as much as if he did. That made an impression on me. But it would take me another decade or two to understand that sharing time with friends who have a passion for fly fishing and the outdoors could sometimes be more important and enjoyable than the fishing itself.

We turned in late that night. The next thing I knew, I awoke on the porch under the quilts. It was early morning, and the first rays of sunlight were streaming through the trees, sending narrow beams of light through the porch screen and down to the old wide-plank flooring just past my bed.

Dad was already back in the cabin with trout in his wicker creel. He had stuffed wet grass in the bottom of the creel to help keep the trout fresh and cool. He used that creel throughout the forties and fifties. I proudly had one just like it, but without the fish.

I remember asking my dad, "What time did you get up?"

He said, "At the first light of day and caught my limit on Jake's Creek before the sun rose."

I thought about Dad getting up so early to catch fish for our breakfast. I was concerned and maybe a bit embarrassed to think that I had overslept and missed the fishing part of the trip.

Dad read my mind and said, "Don't worry, I'll take you out later today when it warms up a bit and let you practice casting, and even fish." That made me feel a lot better.

He and Bill fired up the old gas stove, and soon we were enjoying fried mountain trout, scrambled eggs, grits, and biscuits with honey and jelly. With the sound of the creek in our ears and the first sign of warmth in the air, my worldview of trout fishing was off to a good start.

I didn't catch a trout that day or any other day that year. It didn't matter; I knew I would be casting better when I got older and that I had a good teacher. And while I had no success that year or the next, Dad was patient and would always explain what he was doing and why. Often he would hook a fish and hand me the rod so that I could fight the fish and bring it to the net. So even though I couldn't cast, I felt like I was valued and was learning. And indeed I was.

I don't think there was ever a trout fishing experience in the Smokies that made a bigger impression on me, particularly the memories there on the old cabin porch before I ever so much as touched a fly rod: the sound of the waterfalls, my first taste of fresh-caught trout, the smell of the wild mixed with the wood smoke from the fire and the tobacco aroma from Bill's old pipe. Listening to the fly fishing stories they told. It's seared into my psyche, and it will never leave me.

N
W
E
S

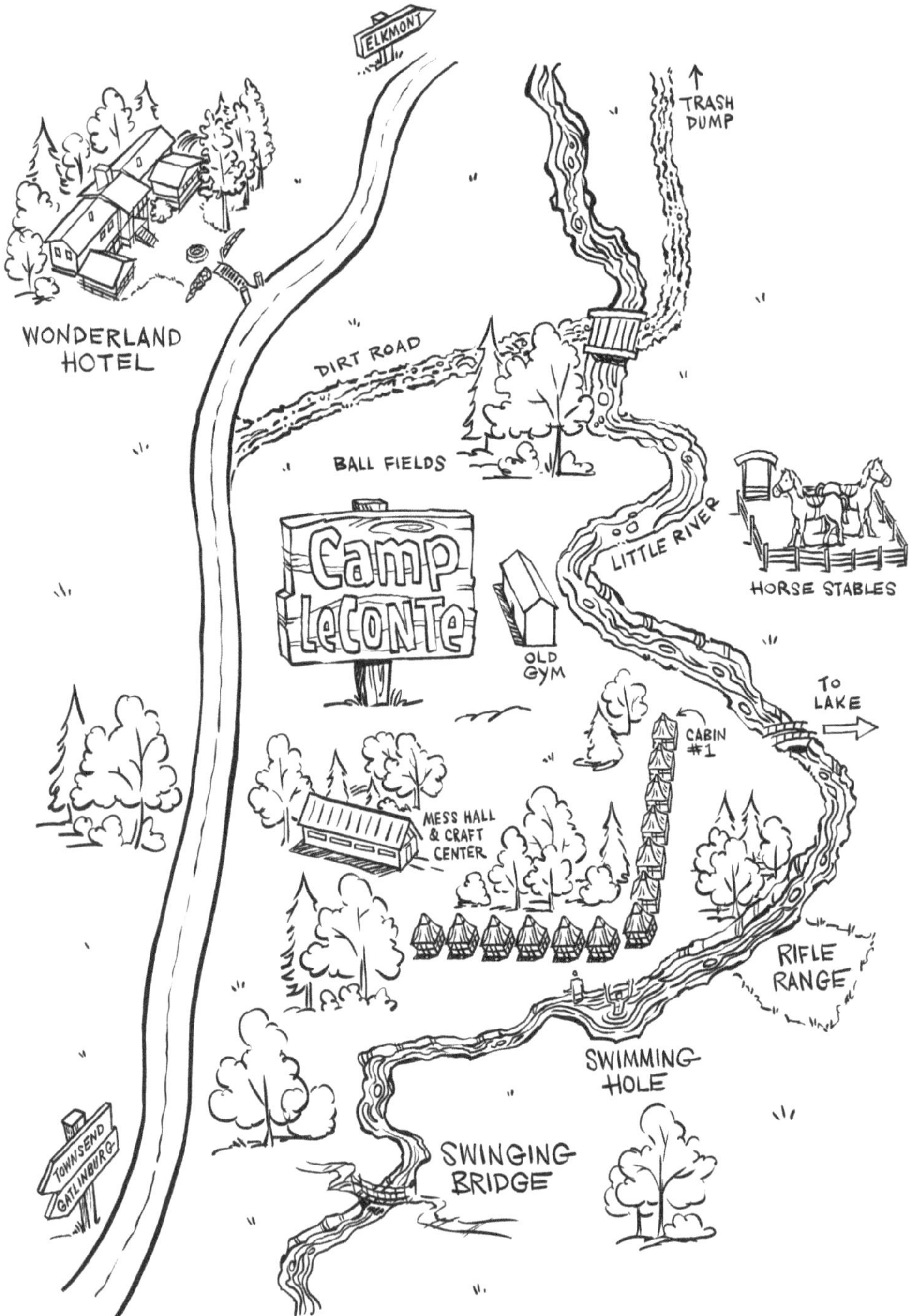
ELKMONT
TRASH DUMP
WONDERLAND HOTEL
DIRT ROAD
BALL FIELDS
Camp LeConte
OLD GYM
LITTLE RIVER
HORSE STABLES
TO LAKE
CABIN #1
MESS HALL & CRAFT CENTER
RIFLE RANGE
SWIMMING HOLE
SWINGING BRIDGE
TOWNSEND
GATLINBURG

A CAMP IN THE SMOKIES, 1949

EARLY THE NEXT SUMMER, at the age of seven, my parents sent me to a summer camp inside the park. Camp LeConte was a boys' camp on the Little River just downstream from Elkmont and the old Wonderland Hotel, where my parents stayed on the weekends when they came to visit.

I remember the hotel's dining room with its fireplace, the checkerboard tablecloths, and the creaky floors. Just outside, there was a picturesque view from a wide porch. It's where the adults gathered in the evenings on swings and rocking chairs and listened to the whip poor wills.

People are always curious how a privately owned camp, a two-story ramshackle hotel, and cabins at Elkmont could even exist inside a national park. It's an interesting story. All three of these places, and others that were there before the park's formation, were grandfathered in when the park was established. The park basically said to the property owners, "Sell to the park. If you do not choose to sell, you can lease from the park. But at the end of the lease, the property goes to the park."

Leases on the cabins at Elkmont were ultimately extended twice by the park. Each extension was twenty years. The leases have all expired now, and the park owns all of it. Unfortunately, many of the cabins were demolished.

The old Wonderland Hotel had opened in 1912. The only access originally was the logging railroad, which Colonel Townsend had built. My grandmother once told me that when she was a little girl, it was about three hours by train all the way from Knoxville to the Wonderland Hotel, unless the engine broke down.

Camp LeConte was just a bend in the road down from the hotel. The summer session at camp was two months long—June and July. There were thirteen cabins for campers, a mess hall with craft center, a lake for swimming, a rickety old basketball gym, a riding stable with trail rides every other day, a rifle range, and a couple of fields marked off for football and baseball.

I was sent there for two months every summer, except for the summer we took a family road trip to Montana, when I attended camp for only a month.

At the age of seven, I found myself in the cabin for the youngest kids, which was cabin number one. Each cabin had eight campers and a counselor. All the counselors were college athletes from SEC (Southeastern Conference) universities. I thought that was pretty cool. The camp was owned by John Barnhill, who was the athletic director at the University of Arkansas. He had played on the offensive line for the University of Tennessee under its famous coach, General Robert Neyland, and was even the head coach at the University of Tennessee during the years when General Neyland was called off to war.

We slept on old military bunk beds—each with two sheets and an army blanket. Each camper had a footlocker for belongings. Every morning there was an inspection. You had to make your bunk military style, and the cabin had to be spotless. It was a pretty tough inspection.

Just like in the military they blew reveille early, and the entire camp formed at attention outside their cabins for roll call. Each cabin's best camper from the previous day was chosen to report. The camp commander stood in the center of the camp under the tall oak trees.

He would call, "Cabin one—report."

The camper reporting would salute and sound off, "Cabin one, all present and accounted for, sir," and so on for each of the thirteen cabins. It was good training and even though I couldn't have known it at the time, I would later head to a military academy in high school.

There were overnight and day hikes to places like Andrews Bald, Charlies Bunion, Silers Bald, Chimney Tops, Mount LeConte, and other famous landmarks throughout the park. In the mornings, we rode horses or swam in the lake or in the river. In the early afternoons, we made crafts, shot 22-caliber rifles at the rifle range, and alternated late afternoons between baseball and football games with the SEC athletes as our coaches.

There were additional football and baseball games in the evenings after dinner. Those were the highlight of our day. Sometimes those after-dinner games got especially competitive because they divided us up by cabin instead of by age group, and bragging rights came into play.

The ball games lasted until dark—and that's when the strangeness happened. To me it was almost paranormal. Sparks of blue light rising in the night, flashing and sparkling like some galaxy a million miles from here.

I remember thinking how marvelous this show was and how it happened almost every night, and I assumed that it must surely happen everywhere. But of course, it didn't happen everywhere. Remarkably, in all the country, it only happened here, in this one place, Elkmont.

Synchronous fireflies. The only species of firefly in America that flashed their lights in unison. Once you saw it, you never forgot.

Fireflies didn't generate a great deal of conversation among the campers. We just sort of took them for granted. However, we did occasionally wonder why they often all flashed at the same time. I overheard some of the counselors say they thought the males might be competing. The cooks in the mess hall were convinced that the male fireflies all flashed together so the females could make better comparisons. I never did find out. Back then, no one seemed to know for sure, and most of us were way too young to give it much thought.

Nevertheless, on some nights the show was unreal. We saw them flash in unison across the ball fields, rolling along slowly like a sparkling wave—on and on and into the darkness of the forest. It seemed strange to me that no two nights were the same. Sometimes they flashed randomly. But mostly they all flashed together, as if they were orchestrated by some mysterious spirit, in a way that we mortal campers were not meant to understand.

Today visitors come by the thousands to the park in hopes of getting parking passes for a fee to view the fireflies. There has even been a lottery drawing and a shuttle service from Gatlinburg. And to think we saw it every evening in the right weather.

One of the world's leading authorities on the firefly is Knoxville's Lynn Faust. Her husband Edgar and his brother Hugh have been lifelong friends of mine.

The Faust family had one of the best cabins in Elkmont. They had purchased the lease from none other than Colonel Townsend himself.

It is only fitting that someone so close to Elkmont became the renowned authority on fireflies. Too bad Lynn wasn't available back then at Camp LeConte to help us understand what in the world those fireflies were thinking.

Each year as I got older, I moved up in cabin number. By the age of nine, I was in cabin number nine. That was the year I made an important discovery. One of the cooks at the mess hall told me there were bears every evening at the trash dump about a half mile up the dirt road on the other side of the river toward Elkmont.

The dump was simply a large depression in the ground where the people from the camp, the hotel, and others brought their trash. Occasionally someone brought a dozer up there and pushed dirt in on top of all of it. I never did learn who that was.

At 6:30 p.m. after dinner in the mess hall, we were dividing by cabins for a baseball game. I suddenly developed a severe "stomachache." The counselors said, "You're too sick to play baseball," and added, "You should report to the infirmary." Of course, when I got to the other side of the ball field, the stomachache mysteriously disappeared, and I didn't turn right for the infirmary. I headed left to the wooden bridge across Little River, and then I took another left up the dirt road along the river and through the tall forest toward the trash dump.

As I approached it, I saw a large bear ambling out of the trees and heading down the grade into the dump. He disappeared, so I crept closer. I was a hundred feet from the dump and could see nothing. It was a little too deep to see the bottom of it from where I was, so I ventured even closer. I got within fifty feet. Still I could see nothing. I knew there was a bear in there. I had seen him enter. What else was in that depression? I had to know.

I sneaked up closer still. I saw the back of a bear facing away from me. And then another, and yet another. And even then, I could only see a small part of the pit. The rest was a mystery.

One of the bears turned, spotted me, let out a gruff "wuff" and ran for the mountainside. This started a chain reaction. Like a stampede, five more bears came bolting, galloping and hurtling out of the dump in full panic, right on the heels of the first one.

I was elated. I had just scared away six full-grown bears. It was quite a rush for a skinny little nine-year-old. I reveled in it. I remember thinking, "Bears aren't so bad. I can scare them easily."

I backed off about twenty feet to some trees and waited. Before long, they came back again, one at a time. They knew I was there obviously. They would each walk to the edge, pause, look at me for a few seconds, and then saunter down into the dump.

Emboldened by my success, and keen to see them run again, I walked directly toward the crater. To my surprise, there was no panicked stampede. No sprint for the trees. Instead, they all looked up at me at the same time.

It's sort of disturbing to see six bears all suddenly look at you simultaneously. I needed time to process that. It wasn't the reaction I expected or wanted. I needed to reassert myself.

I waved my hands high above my head and hollered at the top of my lungs. I was pleased to see the bears run out of the garbage pit, but I noticed they didn't go nearly as far as the first time. I should have taken that as an omen.

Had I lost my mojo? I desperately needed to reestablish the pecking order. I so wanted to see them run full speed again. To feel in control.

It never occurred to me that I only weighed 65 pounds soaking wet. And six garbage-fed adult bears probably averaged 300 pounds apiece. For the benefit of the doubt, call it 250. That's 1,500 pounds of bears versus a 65-pound nine-year-old kid from Knoxville who should have known better.

I took a few steps toward the closest bear, waved my arms and hollered again. This time I gave him top volume.

The last thing I saw was him pinning his ears back as he came charging directly at me.

They say the worst thing you can do in the event of a black bear attack is run. It triggers the bear's primeval instinct to attack and kill. Needless to say, I turned and ran. I fled in sheer terror, instinctively back down the dirt road through the forest toward Camp LeConte.

A bear can outrun a horse in the quarter mile. I ran faster than I had ever run up to that point in my life. And I ran and I ran. For a half mile, I never stopped running until the door of cabin number nine closed behind me. I was that terrified.

I never even looked over my shoulder to see if he followed or how close he got. Of course, he could have had me in the first five yards if he really wanted me. Clearly, he must have preferred garbage to me—which doesn't say much for me.

I was alone in the little cabin for a while, and that pleased me as I wasn't keen to answer questions about where I had been. I opened the squeaky screen door. The sun was getting low. I ran beneath the tall oak trees to the ball field. The counselors were surprised to see me as they thought I was in the infirmary.

They asked nothing as to where I had been, and to my delight, they put me in at my favorite position which was second base. Now the sun touched the mountain to the west. Maybe time for another inning or two before the fireflies appeared and it was too dark to play ball.

I glanced across the open field to the gray wood line where the dark forest began, the place where I had been, and once again I felt the rush, but not the fear, just the excitement of it all—of daring to venture out. It was the first time I had ever felt it: the heightened sense of being more alive. To know something that the other boys did not. To experience what they had not—to feel what they had never felt. And it seemed to me that I was no longer just a regular camper—and would never be again.

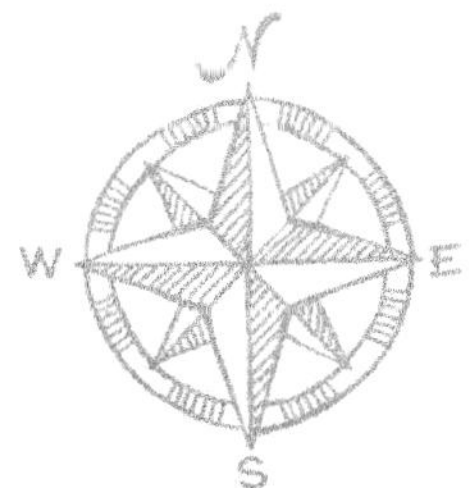
N
W
E
S

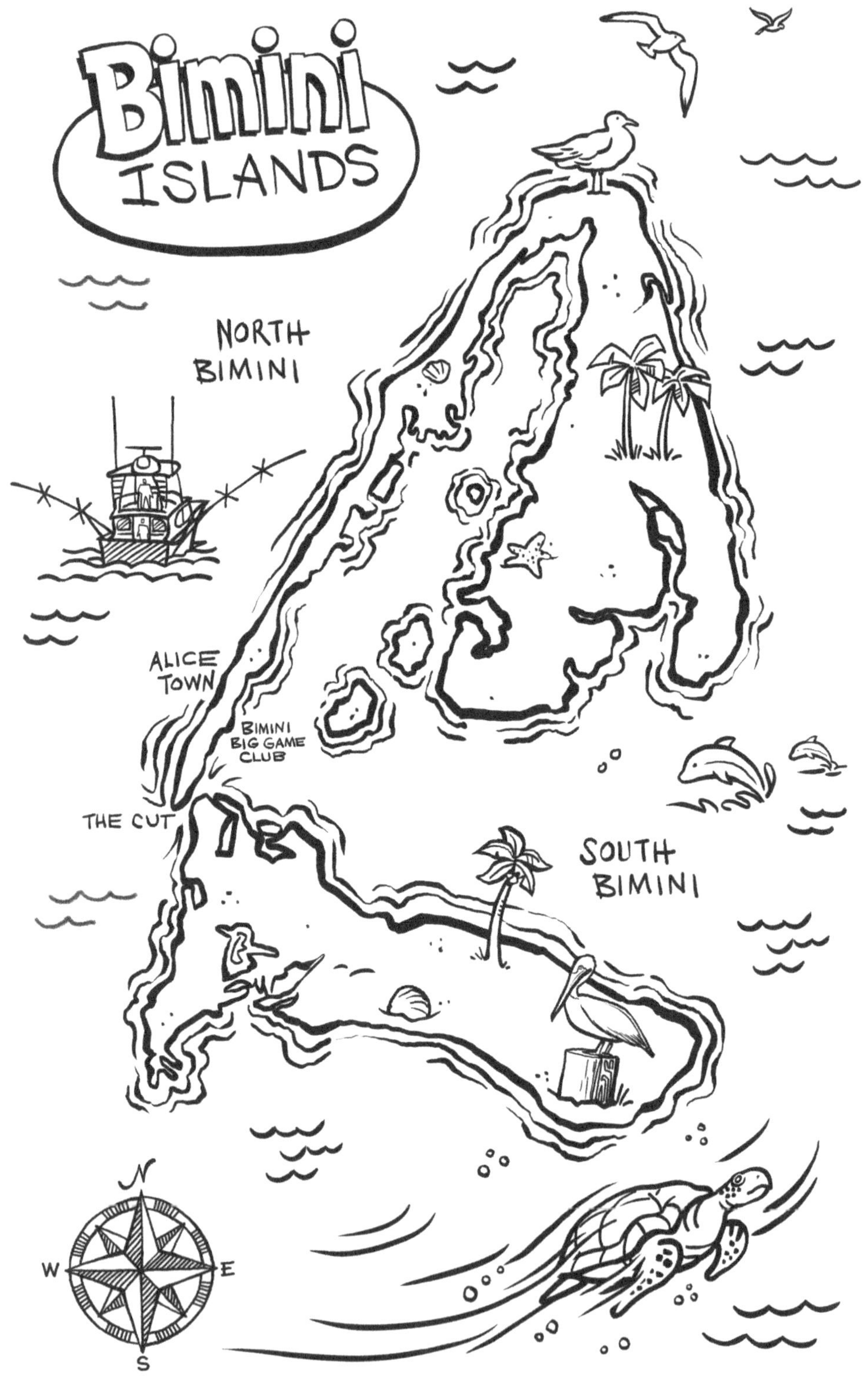
Bimini
ISLANDS
NORTH
BIMINI
ALICE
TOWN
BIMINI
BIG GAME
CLUB
THE CUT
SOUTH
BIMINI
N
W
E
S

THE ISLAND, 1949

I HAD JUST TURNED nine years old, and I was off with my dad on a big adventure.

We crossed the cobalt blue waters of the storied Gulf Stream and Florida Straits. Captain Art Robinson guided the *Vera II* though the cut and pulled into an old wooden dock on the east side of the north island.

It was off the beaten path, but not to a select cadre of celebrities and renowned names in the big-game fishing world. It had been a favorite hangout of people like legendary angler and writer Zane Grey and his captain, Tommy Gifford, who invented the outriggers now used on nearly every deep-sea, sport fishing boat in the world. The inner circle included recluse Howard Hughes; retailer-turned-big-game-fisherman Michael Lerner, the founder of the International Game Fish Association (IGFA); Ernest Hemingway; and other well-known big-game anglers of the day.

Of course, at the age of nine, I was unaware of any of this, but it was here in Bimini, as well as in Key West and Havana, that Hemingway wrote amazing stories and fought big marlin and giant tuna season after season, becoming one of the most celebrated fishermen in big-game fishing history. In the very year of this story, it was here in Bimini that he began writing *Islands in the Stream,* even though it was not published until 1970, well after his death.

There were three or four sport fishing boats tied to the rickety old dock that evening. We pulled into a slip next to a boat called the *Pilar*. To me, it meant nothing. Just another boat.

Of course, it was just the most famous sport fishing boat in all the world.

It was a thirty-eight-foot Wheeler with a single engine. During World War II, Hemingway had used the *Pilar* to help hunt for German U-boats, and at that time, the *Pilar* had been outfitted with direction-finding equipment in the cockpit, a Thompson submachine gun on the bow, and hand grenades and ammunitions stored below deck.

Hemingway had modified the *Pilar*, and it continued to evolve over time as he added a rudimentary fly bridge, a fighting chair, and a roller across the transom to help pull the larger big-game fish into the boat.

As a nine-year-old, I was more excited about fishing in the ocean and clueless to all that and to most everything else. I just knew I liked fishing trips with my dad. For me, this trip to Bimini was a very big deal, the biggest thing in my life so far.

Once the boat was tied and secured, I was left on the dock while my dad and Art headed off to the Bimini Big Game Club for dinner. They said, "You are welcome to come along, but there will likely be heavy drinking in the club. Maybe it's best if you stay here at the boat, and we'll bring you some food." And off they went.

I had one of Dad's bass fishing rods, and I was happy just sitting on the weathered dock and fishing for the small sheepshead that I could see clearly in the water below. A Bahamian boy my age came by. We chatted boy talk. He kept eyeing my knife in the sheath on my belt. And I kept looking at the huge conch shell he was carrying. I had never seen a shell that big. Mom and my sister, Nancy, and I had hunted for shells on the beaches of Pompano and Fort Lauderdale in the early mornings, and the few shells we had found were very tiny. I desperately wanted to show my mom that huge shell.

Pretty soon a trade was made. I gave him the small knife and he gave me the conch shell. I asked him to show me where the Bimini Big Game Club was. I wanted Dad to see the shell.

It wasn't hard to find. Loud Bahamian music was blaring from within and drifting across the narrow island. As I approached the club, I heard the voices of adults, men and women, laughing and talking loudly. I wished I was older so I could understand more about what they were saying.

On tiptoes I peered through a window. I saw a room crowded with tables, lots of food, drinks, cigar smoke, and people. I spied my dad across the room at a table with Art and some people Art seemed to know.

I timidly ventured from the window to the door and into the club. The energy, the noise, the strange and loud music, and the feeling that the people in this room all knew each other was overwhelming. My Dad pulled up a chair for me. Art and dad ordered me a bottled Coke and some fried conch for dinner.

I showed them my shell, and Art said, "The conch you are eating came from that." I was amazed. I didn't even know food came from seashells. It made the dinner all the more remarkable.

There was a longer table right next to us with ten or twelve people who obviously all knew each other and were having a very boisterous good time. There was a striking gray-haired, heavyset man with a trimmed beard sitting at the end of the table right next to me. I could have touched him. He was that close. He seemed to be the one holding court, telling most of the jokes, recounting the fishing stories, and commanding most of the attention not only from the people at his table but from others in the room.

My dad leaned over and whispered to me, "That man next to you. Do you know who he is?"

I answered, "No."

Dad said, "That is Ernest Hemingway."

I then loudly blurted out one of the most profound statements of my life. "Who is Ernest Hemingway?"

Dad patiently explained who he was and that he was one of the most famous people on Earth, if not the most famous. I nearly froze with intimidation. I was almost afraid to even look at Hemingway after that, even though he was right there next to me. But I listened to the stories and the way the room responded when he told a joke and how he held them spellbound. The tales he told of fighting big marlin and tuna. I couldn't understand much of it, but it excited me, and I got the drift. The thought occurred to me that maybe one day I could sit at the end of a long table like that, tell stories of fighting giant fish, and that people might actually listen to me.

We slept on the boat. In my head were dreams of the fishing tales I had heard, and early the next morning we went to sea.

There is a famous saying in the fishing world—"You fish until the fish stop biting." As it turns out, that is true everywhere in the world except for one place: the place where we were headed.

Out in the ocean, and well out from Bimini, there were a couple of big rocks that jutted abruptly from the deep blue Gulf Stream water. They strangely resembled a large and a small chicken. These unusual rocks, known as Big Chicken and Little Chicken, and the shallows around them, were immediately adjacent to deeper water, which Art said "makes it a great place to fish."

Unfortunately, these rocks were also used by the US Air Force out of their base in Homestead, Florida, for target practice with their fighter planes and bombers. There was an informal agreement between Bimini, the boat captains, and the US Air Force: fishing was allowed at the Big and Little Chickens only if arranged and approved by the air force and the Bahamian government in advance, in which case the Air Force would hold off strafing runs while you were fishing.

It had only been four years since the end of World War II. To me, armies and warplanes were bigger than life. So, there I was, a nine-year-old, standing on the deck of the *Vera II*, listening to Art on a radio call with a real US Air Force squadron commander, a couple of US fighter pilots, and the Bahamian government. My mind was about to explode. Suddenly a fishing trip to Bimini had turned epic. I was wide-eyed listening to Art communicating with the fighter squadron commander. After about ten minutes of negotiation, the Air Force agreed to hold off bombing runs on Big Chicken and Little Chicken until 11 a.m., and the government official granted us permission to fish.

As we approached the Big Chicken, the water surrounding it was turquoise, clear, and calm. It was about 9 a.m. Art cut back on the throttles, and instantly we were looking down over the gunnels of the boat at not one or two, but hundreds of huge amberjack. I had never seen anything like it—like whales in a swimming pool.

To a boy who grew up fishing for bluegill and crappie in Tennessee, this was mind-blowing. The water was so clear. It was almost like the swimming pools that I was accustomed to back home, except this pool was full of thirty-five- to sixty-pound fish.

The next two hours were the most phenomenal moments of my life to that point in time. There was seldom a minute my dad and I didn't have an amberjack on our lines. Usually, we were fighting doubles. It was unlike anything I had ever dreamed of. How could I? I didn't even know fishing like that existed. And to make it even better, I was experiencing this with the person I admired most in my life—my dad.

Minutes seemed like seconds to me. Time flew by, and I did not want it to end. We hooked fish after fish. On some level I knew it was me that was really getting hooked, and that somehow fishing for large fish in oceans would become part of my future life. It had to.

I was busy fighting a fish, as was Dad, when from the corner of my eye I saw something unexpected and alarming. Art suddenly pulled a large knife from the sheath on his belt, advanced toward us, and with one purposeful and swift motion, cut both of our lines. All he said was, "It's time to leave." He quickly cranked up the engines on the old *Vera II* and we roared away from the Chickens.

I was heartbroken that we were leaving. It had been fishing on a scale I did not know existed. I couldn't understand the hurry to leave something this momentous.

Suddenly there was a roar, louder than anything I had ever heard—like a thousand jackhammers. A US Air Force fighter plane came roaring in low and fast right over the top of our boat, nearly at water level, just barely clearing the tops of our outriggers. Its guns were blazing as it sent a spray of machine gun fire into Big Chicken.

Art just smiled. It was 11 a.m.

Big fish, Air Force fighter planes, machine guns, famous author—I somehow knew that this trip to Bimini had changed me, but I did not yet know how or what it meant. I just knew that someday, I wanted to fish the oceans of the world and maybe even be in a war where there were armies and fighter planes.

Be careful what you wish for.

A ROGUE WAVE, 1950

I WAS TEN YEARS OLD. The family had packed the old woody station wagon for a beach vacation to the southeast Florida coast.

It was a two-day drive, and to a ten-year-old kid anxious to go fishing in the ocean again, I thought the road trip would never end. I passed the time by reading parts of two books my teacher had assigned for summer reading: one was about a US Army Ranger battalion that scaled the cliffs at Normandy and the other was a story of the paratroopers who jumped into the Netherlands in the Battle of the Rhine. I read and reread those chapters again and again—and I wondered if one day I might be a soldier, a paratrooper, or a Ranger. I hoped so. It was an exciting thought, but just a dream.

It was a great vacation, and we alternated days between the beach in Fort Lauderdale and deep-sea fishing out of Pompano. We had great weather for our first day on the beach. But the next day turned windy. Maybe too windy to venture out on the ocean. The charter boats back then were small—not even half the size they are today. So we had a discussion with our captain, Art Robinson, and after much debate, we decided to venture out anyway.

The seas were rough and high, like nothing I had ever seen, but I liked the waves—to see them coming, to feel the lift of the boat and the fast drop as the boat slid into the giant trough. I was up on the bridge with Art, and I asked him about the waves. And he said, "They are born in the tropical winds, and some of them travel many miles as they gather speed and momentum."

He looked at the horizon, beyond the waves, and what he said stuck with me for the rest of my life. He said, "Waves are sort of like people. No two of them are exactly alike. Some are big. Some are small. Some are friendly, others not so much. Some are weak, and some are unpredictable." Then he paused, "And even violent. There aren't many like that; maybe once every few years. Those are the ones we call the rogue waves, the ones you watch out for—so they don't catch you by surprise."

I thought about his words, and as I felt the strong wind, I wished that one day I could be a boat captain and fish the oceans. There was adventure here, I was sure of it, and fish that were big and fast and strong. And while I was way too small to land most of them, I hoped that one day I would grow strong and understand the sea—and be like him.

Later that morning when the wind blew even stronger, I went below deck to fix a sandwich. I had started back up to the deck above, and that is when it hit—when the whole boat rose into the air and lurched violently to the side. It lifted higher than I could have imagined, and it caught me by surprise. The next thing I knew, I was thrown off my feet to the other side of the cabin, and I hit my head sharply against a wood cabinet. The blow hurt as I made my way up to the deck, which was full of water. My mom and dad were shocked. Art yelled down from the bridge, "That was a big one." I wondered if it was the rogue wave.

For the next hour or so I did not feel right. My head was throbbing, I began to feel nauseous, and I started vomiting. Everyone thought I was seasick. I knew I wasn't seasick because I had fished on the ocean for several years and had never been seasick. I loved the waves. I argued the point but was overruled.

When we returned to the hotel that evening, I worsened. My mom and dad became alarmed.

They somehow found a doctor in Fort Lauderdale that evening who examined me. There were terrible outbreaks of polio in those days. The doctor was concerned because the back of my neck as well as my ankles and lower back had less than normal range of motion. The doctor arranged for Mom and Dad to take me to the Coral Gables Children's Hospital in Miami for testing. I don't remember much about the drive to Miami that night because I was so sick.

The doctors at the hospital told my parents they suspected polio. They said the only way to know for sure would be a spinal tap. I recall it being painful, and since I was already sick, it made the nausea worse.

The results were positive. I had polio. It was a highly contagious disease that attacked the grey matter in the spinal cord, which affected the central nervous system and caused paralysis. I was rushed to the quarantined polio ward, which was a large room about the size of a basketball gym floor with lots of beds and patients. They allowed no visitors, not even family.

There were no vaccines or antibiotics back then for polio. The doctors told me, "Paralysis doesn't show up until the second week." And they added, "The paralysis will be worse the more you move around during the first week." And they told me, "Lie still and do not move."

For some polio patients, there were no symptoms during the first week. As a result, there was no reason to seek treatment. Suddenly a person became sick and paralyzed at the beginning of the second week after being infected, often resulting in permanent disability or death.

Immobilization was the primary treatment, particularly during the first week after infection. Lying completely still is difficult for adults, but it is worse for a ten-year-old kid. The nurses kept warning me that the more I moved around in my bed, the worse the paralysis would be when it hit. On some level I believed them, but it was a hard thing to grasp as a child. When they used the term "move around," they literally meant not fidgeting, squirming, turning over in bed, sitting up, tossing, or twisting. It was a concept I could grasp on one level, but the practicality of the matter was that it was nearly impossible for a kid to do that—much less for weeks at a time.

Other than immobilization, the only other treatment was to wrap the patient in heat. Three times a day, the nurses wrapped me in very hot towels for thirty minutes. As the steaming towels began to cool, they applied new hot towels. The hope was that the heat activated the body's immune system to help find and fight the poliovirus hiding inside the spinal cord and other parts of the skeletal system.

Every day a group of doctors came by my bed and examined me. Like the nurses, they told me to "always lie still." I even overheard one of their private conversations in which the nurses told the doctors that they were worried about me because of my constant shifting around in the bed. Later that morning, one of those nurses threatened to strap me down if I kept moving. I only had to be strapped down once. It was not pleasant. I guess looking back on it, I understood what they were telling me, and I tried to comply, but deep down it was hard for me to believe I would be paralyzed.

Very close to my bed were three patients in huge iron lung breathing machines, which made ominous swooshing sounds that terrified me. I thought the iron lung machines looked like torture chambers. Patients were placed inside the iron lungs with only their head showing. A bellows sucked the air in and out, forcing their chests up and down to help them breathe.

Seeing them suffer like that gave me the worst possible fear. One of the nurses told me I was headed to an iron lung if I didn't lie still. I'm glad she did that. If I could see her again, I would thank her. I think that warning, more than anything else I was told, made me stop moving around in the bed.

It was a scary situation for a ten-year-old, away from home, unable to see my parents, in a large room with forty or fifty other polio victims, and being told that if I moved, I would be paralyzed. I knew my parents loved me so much and they must be terribly worried. Of course, I had no contact with them, even by phone. There were no phones in the polio ward.

I hoped I would be okay. My parents had always been there for me. I imagined them talking to the doctors about inventing some new cure that would save me just in the nick of time.

I entered the dreaded second week. About once every hour or two I felt a severe and sharp, stabbing pain that would usually start in my lower spine and work upward toward my neck. It caused me to arch my back and sometimes I cried out until it subsided. I felt the ache deep into my legs. Each episode of that lasted a minute or more. Maybe I was too young to fully comprehend it, but at the age of ten and alone in that polio ward with no friends or family, I was deep into one of the most dangerous fights in my life.

The stabbing pains continued, day after day. And then, after three or four weeks and as if by some miracle, the pains slowly began to subside. One day two doctors came to my bed and smiled at me. They said it was looking hopeful that there would be no paralysis. The two modalities of immobilization and hot towels continued.

As it turned out, I was one of the fortunate ones. I made it through the dreaded early weeks without paralysis. I benefited from an early diagnosis and from being immobilized in week one.

After five weeks in the polio ward, I was feeling much better. I was desperate to get out of that bed and out of the hospital. I hadn't seen my parents, or anyone I knew, the whole time, and I was homesick for my life in Knoxville and my friends. But the doctors wanted to be sure I was ready, and disease free, so I would not go out and infect anyone else.

I also remember the doctors and nurses telling me that if they discharged me before I was totally disease free, the chances were great that if I walked or ran or moved around in any manner, the polio would return, and I would be paralyzed. The doctors kept me there an extra week as a precaution.

Finally, at the end of six weeks, three doctors came by my bed to evaluate me for discharge. The tests were primitive, mostly range of motion and flexibility. The primary test was when they checked to see if I could touch my chin to my chest. If polio was still present, that would have been impossible. I passed their tests.

The day I was to be discharged, I was told not to run, play outdoors with other kids, or exercise in any way for a month. I think for most children, that would have been a real blow. But by this time, I realized that I was one of the few lucky ones. I had made it through the ordeal with no permanent disability. When I returned to Knoxville, I was kept in the house to be sure I followed the doctors' orders.

At last, after six weeks in the hospital and a month under house arrest, I was free. I noticed that when I ran and played football with the kids in the neighborhood, I was not as fast as I used to be. But slowly over the next six months I got stronger, my speed returned, and I became my old self.

It was a great feeling to be free, and a kid again, and cured. Others were not so fortunate.

Following my bout with polio, and over the next twelve months, of the 58,000 adults and children in the United States who contracted the disease, 3,200 died (5.5 percent) and 21,000 (36 percent) were paralyzed permanently. For some, the disease spread to their brain and to the muscles in their chest and they couldn't breathe. Doctors were able to save some of them with the iron lung, which provided artificial respiration for a while. Some of the patients were then able to breathe on their own. Others were not so fortunate and died.

I heard the doctors tell my parents how fortunate it was that I had hit my head against the cabinet in the boat in the rough sea, because it had triggered earlier onset of the disease. It meant I got tested, diagnosed, and treated earlier than most. Therefore, I was immobilized during that critical first week.

That one big rogue wave had changed my life.

I've often wondered about that wave—and the tropical winds that started it, and the distance the wave might have traveled—and the likelihood of me being in its path at just the right time. The remaining stories in this book would not have been possible had I been one of the less fortunate ones. Perhaps God sent the rogue wave in my direction at just the right moment in time, but I was too young to contemplate such a thing.

THE MONSTER AT THE SWINGING BRIDGE, 1952

IT HAPPENED ONE Sunday afternoon in the late summer. I was finally in cabin thirteen at Camp LeConte with the older boys.

Afternoon was free time, and some of the cabins challenged other cabins to football. I really liked football, but that day, I had other plans: I was heading to the river.

The Little River ran through camp. It was mostly shallow and fast with a few deep holes. One was the swimming hole. Another was the swinging bridge hole. It was downstream from the kitchen and mess hall and just before the river ran off the camp property.

Lately, after meals in the mess hall, I had been taking bread from the dinner table to the swinging bridge hole, making doughballs and feeding the trout. From up high on the swinging bridge, I looked deep into the clear water. For me it was fascinating to see the ten- and twelve-inch rainbows feeding on this "manna from Heaven."

Then one day I spotted something else. I noticed that deeper, in the darker water near the very bottom, the white doughballs just disappeared. Occasionally I thought I saw something big. A long, dark shadow that seemed to move? Day after day I marveled at this phenomenon, but I couldn't figure out what it was. Was it a monster trout? Or was it just my eyes playing tricks due to the reduced visibility so deep in the pool?

After a week of this mystery, I decided to solve the riddle. I pulled my fly rod case and reel from the trunk at the foot of my bunk and rigged my leader with a hook and split shot for weight. With a few slices of Kern's white bread in my pocket, I headed for the swinging bridge.

For a while, I fed the fish from up on the bridge. The twelve-inch rainbow was there along with some smaller trout. I lowered the doughball on the hook into the deep pool below and immediately hooked up with the twelve-incher, hoisting it up on the bridge where it flopped around without breaking my leader. With my pocketknife, I cleaned the trout as Dad had taught me. I took it to the head chef at the mess hall with a plea for her to store it in their refrigerator and "please cook it when my parents come from Knoxville to visit this evening."

I was proud of that fish and wanted to show it off to Dad even though I hadn't caught it on a real fly.

The head chef lady said, "I'm not allowed to cook for the parents, but I'll be happy to save it for your dad in the refrigerator so he can take it home to Knoxville." That was good enough for me.

It was still early afternoon, and I went back to the swinging bridge. Once more, I fed the fish, but I never saw the shadow or anything else unusual. Finally I tired of pitching doughballs into this pool with only a few very small trout. I walked back across the swinging bridge to the camp side of the river and went to the smaller pool in swifter water just above the bridge. Here the deepest water was on the far side of the pool.

I tossed a doughball over where the water had cut a run by a boulder. The doughball sank a foot or so as it passed the boulder, and I saw the white of the doughball disappear suddenly like something larger might have taken it.

I then tried a doughball rig on my fly rod to no avail. The fast current in the middle of the stream would catch my line and pull the doughball along faster than the current as it came by the undercut rock. Apparently, whatever was there didn't like that, and the game ended. I wondered silently if the big fish in the swinging bridge hole sometimes came up here to rest in the shadow of the undercut rock.

I thought to myself, "What could I cast over there that would look natural when the fast current in the middle of the stream catches my line and pulls it faster than the current by the rock?" I had not brought my fly box with me to the stream, thinking I wouldn't need it. But it was close by in cabin thirteen, so I returned there for it.

One of the campers in the cabin had seen me take the trout to the mess hall, and he was all excited. He wanted to come along if I went back to the river to fish. His name was Cliff, and he was from New Orleans. So Cliff and I walked back to the hole above the swinging bridge.

I dug around in the little fly box. There were the wonderfully muted flies that Dad had given me to take to camp. Flies like the Gorgeous George nymph and the Tellico nymph tied by his friend Eddie George. The dry flies ran the gamut from Olives to Quill Gordons, Hendricksons, Pale Duns, Gremlins, Royal Coachman, and Cahills. But among all those flies, there was not one that would have looked natural being pulled faster than the surrounding current when the fast water in the middle of the river caught my fly line.

Except for maybe one. It was a fly that Dad's friend Bill Newman had given to me one evening as we sat on the porch of his little cabin on Jake's Creek. It was called a Brown Wasp, and you pulled it behind a tiny narrow blade copper spinner designed to cast on a fly rod.

He had said, "Son, someday you'll get yourself in a place where there is a big fish and no way to get a drag-free float to him. When that happens, just throw this at him and get it moving the second it hits the water. I don't know why, but the bigger fish in this river can't resist the Brown Wasp when they are feeding."

And so it was, that June afternoon in 1952, that I tied on the Brown Wasp, and with Cliff from New Orleans behind me just to keep him from spooking the fish, I cast to the undercut rock. Immediately the faster midstream current caught my fly line and got the fly moving.

The little copper spinner danced and spun enticingly in front of the Wasp. It all looked so right. In my mind, I could feel it. I imagined the big fish watching from beneath the undercut and moving to it with the swirl of a gigantic take. But in front of me on the stream, there was nothing. No reaction or response of any kind that I could see. Had the fish moved? Had it seen us and spooked? Did he even want a Brown Wasp? Was there even a monster fish in this hole?

I cast again, and all these questions were answered in a heartbeat. Like a phantom, he just appeared. I never saw him come from the undercut. I never saw him rise upward from the bottom. Suddenly he was there, inches from the fly and tracking it across the current, leaving a wake so large it disrupted the flow of the stream. I saw him open his enormous mouth as he moved closer to the Wasp from behind. The whole inside of his mouth was creamy white like cotton candy and looked as big as a cantaloupe. I watched in disbelief as the huge jaws crushed downward on the little fly, and I pulled back to set the hook.

I felt the weight of this massive fish as he surged and turned in the current. In the sunlight, I could see his rainbow-hued side and every spot on his immense body. What I could also see was that the fly was not well back in his mouth. It was way out on the tip of his kype. He was barely hooked.

I heard Cliff screaming hysterically behind me. He had never so much as trout fished in his life, but even he could see this fish was a giant and knew that something epic was happening.

The huge rainbow swirled sideways in the fast current, shook his enormous head from side to side twice, and the little hook came off the end of his kype. There was a massive explosion in the current, and he was gone.

I was shaking nearly uncontrollably. In my excitement, I had seen this big fish coming, and with a case of the jitters, I had set the hook too soon.

Now, years later, I know the lesson well. But as a kid, I didn't understand that if I had just waited a second or two until the great fish turned before I pulled back, I would have hooked him in the corner of the mouth and the hook would not have come out.

I was just a kid, so undoubtedly, I would have lost this monster fish anyway, even if hooked solidly. A rainbow trout of that size, and me with so little experience—the chances of me landing him were nearly zero. But I would have loved to have tried.

The fish was in the twenty-eight-inch range and several orders of magnitude larger than any rainbow I would ever see in the Smokies again in my lifetime. To this day, I have no adequate words to describe it—no idea how a rainbow trout got that big in such a small stream. His brothers and sisters usually topped out in those waters at twelve to seventeen inches, and only if they lived long enough without first getting caught by someone.

Mom and Dad visited me that evening and we took a walk along the river to the swinging bridge. I showed them where it had happened. Dad said he had never caught a rainbow over eighteen inches in all the years he fished the Smokies.

Before they left for Knoxville, we walked to the mess hall and the head chef gave Dad the twelve-incher I had caught along with a little bag of ice to keep it cool for the trip home.

I went back to the hole above the swinging bridge many times that summer and through the years that followed, but I never saw that fish again. My biggest fish ever from the park, two actually, were browns, one from Little River just above Elkmont and one from the headwaters of Deep Creek on the North Carolina side. But no fish I ever saw anywhere in all the years and all the miles of streams I fished in the park ever came close to the monster of the swinging bridge hole.

PART TWO

The Formative Years

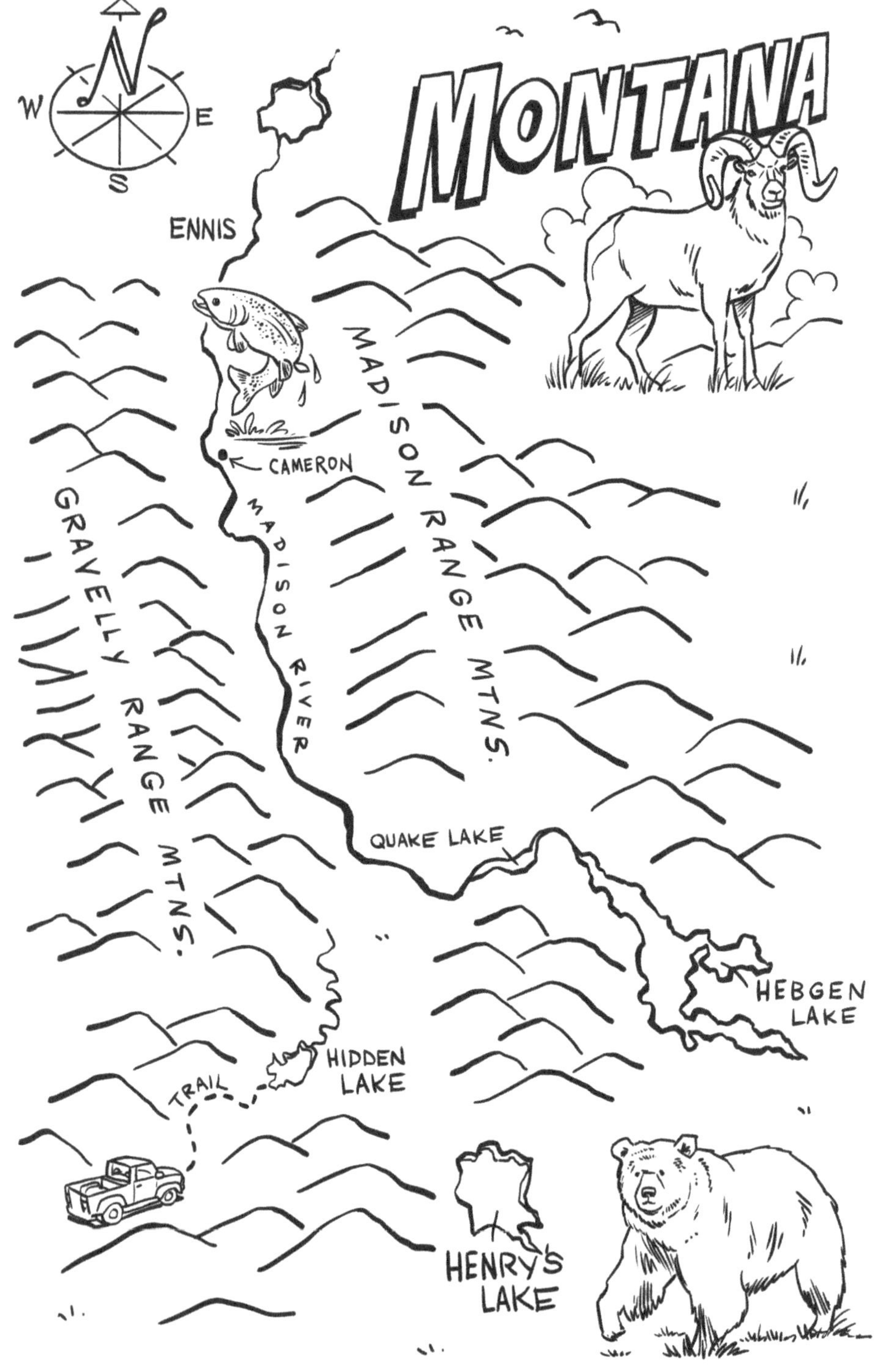
N
W
E
S
MONTANA
ENNIS
CAMERON
MADISON RANGE MTNS.
MADISON RIVER
GRAVELLY RANGE MTNS.
QUAKE LAKE
HEBGEN LAKE
HIDDEN LAKE
TRAIL
HENRY'S LAKE

MONTANA, 1953

MY LOVE AFFAIR with Montana started in the summer of 1953. My parents took my sister, Nancy, and me on the obligatory western road trip. For our family, it was a big deal: it was the first and only time my dad ever took a vacation of more than a week.

We packed the station wagon and drove off into the sunset, down through Arkansas, Louisiana, Texas, New Mexico, Arizona, and up through California, then on through Oregon, Washington, and up to Banff National Park in Canada. Then we came down through Glacier National Park and finally into the heart of Montana.

We stopped at all the usual places, the canyons and national parks. We marveled at the wildlife and western vistas we are all so proud of. All of it was beautiful, but not something that I seemed to be a part of.

Then something happened in a most unusual place.

It was about fifty miles north of West Yellowstone—a little town we stopped in called Cameron, Montana. The population was six, and that was generous. Leonard and Janet McAtee ran the general store and gas station, which doubled as the post office. Next door was Jane's Café. Just past that were six fishing cabins and the Blue Moon Saloon. The whole town was on one side of the road, with nothing but sage and cottonwood trees on the other side.

We met up with friends of my dad, John and his wife, Pat O'Neall, and we all drove into Ennis for dinner. Afterward, we spent the night in two of the little fishing cabins in Cameron. The next morning, we went fishing. It was July 12, 1953.

John said we would be fishing something called the Salmonfly Hatch. He told us that all the big fish in the river would be very close to the banks, feeding on the salmon flies that fell or were blown off the small western willow trees. He gave my dad and me some large flies he had tied. He called them Sofa Pillows. I couldn't believe how large they were in comparison to the little flies we used back in the Smokies.

We drove Highway 287 South, which is upstream, toward West Yellowstone. John explained that we had almost missed the hatch, which moves upstream every day. He said, "There are still a few flies remaining at a place farther up the river called the Cow Camp."

We arrived there about 9:30 a.m. Pat hiked far downstream to fish back up to the car. John took Dad and me down to the river to show us the technique on fishing the salmon flies.

John began his instructions with my dad, and after a few minutes, he took me upstream around the bend and asked if I would mind if he showed me how to do it, and then I would be on my own.

That sounded good to me.

He said, "First of all, don't walk right up to the riverbank because the biggest fish in the river may be right there in shallow water next to the bank, waiting on a salmon fly, and you'll spook him." He said, "Stay back about six or eight feet from the river and make your first cast as close to the bank as possible."

He demonstrated and immediately hooked a spry rainbow that took off down the river. John followed it, around the bend and out of sight.

My one lesson had ended quickly.

Let me explain that the largest trout I had ever caught in my life up to this point on a fly was about thirteen inches long. It was from Abrams Creek in the Great Smokies in East Tennessee, and I had been so proud of how large it was. I was thinking to myself how exciting it was that the big fish in this river were close to the bank, and it occurred to me that I might hook one without even wading out into the water.

It was one of those perfect bluebird, big-sky Montana days. There was no wind. The temperature was warming, and the scent of sage was in the air. The river was big. I didn't even know that streams that big and fast held trout. I was in a state of wonderment, and the river seemed mysterious, but in a good way. I was ready to begin solving the mystery.

I tried to do what John had said. My first cast was close to the bank. My next cast was a few feet farther out.

Seven decades later, I still remember my third cast. The fly landed upstream, about six feet from the bank. It was bouncing along the ripple, riding beautifully on the moving surface. Beneath my fly, I saw reflections of wonderful rust-colored stones the size of a fist. Something else, too.

The bottom of the river seemed to move. Was I seeing things? How could that even happen?

I cast again to the same spot. The salmon fly landed and started its ride. When it got near the place where the strangeness had occurred, the bottom of the river moved again and became a huge fish, which rose up and engulfed the fly.

I was completely dumbfounded but in tune enough to know that this fish was enormous. As I pulled back to set the hook, the trout looked to me to be larger than the ones on the tackle shop walls I had seen that morning in Ennis.

I didn't know what would happen next, but that fish was about to give me a lesson, and part of that lesson concerned my fishing equipment. Back then Dad and I were still using the old South Bend automatic fly reels and glass Shakespeare Wonderods. The biggest downside of the automatic fly reel, and there were many, was that the more line the fish pulled off the reel, the tighter the drag would get. It was designed for the small creeks back east where most of the fish weren't large enough to pull line out.

Here on the Madison River in 1953, that was not the case. The instant this fish knew he was hooked, he went on a wild tear across the river.

It is worth mentioning here that the Madison is wide—very wide. It is also moderately shallow, clear, and for the most part, the current is very swift.

I was completely unprepared as this monster fish tore across the current, accelerating on a mad rush for the other side of the wide channel near an island. I was still standing about six feet back from the bank with no knowledge of how to fight a big fish.

He jerked the rod out of my hands. The rod went sailing through the air and landed on the grass and was being pulled rapidly toward the river. I ran and grabbed the rod just as it slid across the rocks and into the water, which probably saved me tremendous embarrassment later.

By now the fish had gained full speed, the line was sizzling off the reel and cutting across the river, and the drag was getting tighter and tighter. It was a blur: I couldn't think fast enough to move the lever on the reel that would reduce the drag pressure.

When the fish reached the island on the other side, the drag was at full force, which meant no more line was coming off that reel regardless of how hard the fish pulled.

By now the heavy current in the middle of the river had caught the fly line, creating even more pull on the fish. Between the current, the drag, the big fish, and me, something had to give.

Then the big fish jumped near the far bank. It's hard for a fish that large to jump. It was more of a slow lunge skyward. The sun was behind me and shining directly on the fish. He was perfectly illuminated in gold, pink, and luminescent brown. Forty yards away was a little twelve-year-old boy from Tennessee who had never been to Montana. He was trying to grow up and prove he was a man by landing this monstrous fish.

Then the line went slack. The fish crashed back down into the water. The hook had pulled out. In fact, the hook had straightened out completely. The fish was gone, and the Madison was back in equilibrium—but my life was not.

A few minutes earlier, I had been a well-rounded boy on a family vacation. In a matter of seconds, Montana and fly fishing and big trout and big skies and every other big thing out here had me in its grip. The rest of that week only reinforced it.

For the next six days, the boy from Tennessee saw broad Montana valleys framed by snow-capped peaks all around. In every one of those valleys there was a magnificent river running swift and cold. In each of those rivers there were trout—and the entire land seemed like magic just waiting to be discovered.

I had grown up with the Lone Ranger, Gene Autry, Roy Rogers, and Cisco Kid radio programs. Here in the Madison River Valley, I saw real Montana cowboys who looked the part. They ran cattle in the valley in the winters and the high country in the summers with fall roundups just as in the old days.

There were still a few ranchers alive out here in their eighties who had been little kids when Custer was wiped out at the Battle of the Little Bighorn. And there were a few older patriarchs still alive on some of those ranches whose fathers had settled the valley, fought for water rights, saw the advent of fencing, and witnessed the death of the open range. The golden age of the cowboy was long gone, but not in their minds. And not in my twelve-year-old mind, either.

There were very few fly fishing guides in the early fifties in Montana. I don't even recall seeing or meeting one. No drift boats, either. There weren't many anglers at all, for that matter, and those few just waded the rivers. Guides weren't necessary. There were just miles and miles of trout streams and not many people fishing.

It was too much. I felt as if I had been hit by an earthquake, shaken by the revelation that life is much larger than I had thought. The drive from Montana to Tennessee was a long one for our family, but not for me as I sat in the back seat of the old station wagon and relived, over and over in my mind, each new and exciting moment in Montana. I could not have begun to articulate it at that age, but on some level, I was changed, and I knew that no matter where my future life took me, my soul would forever be in the state of Montana.

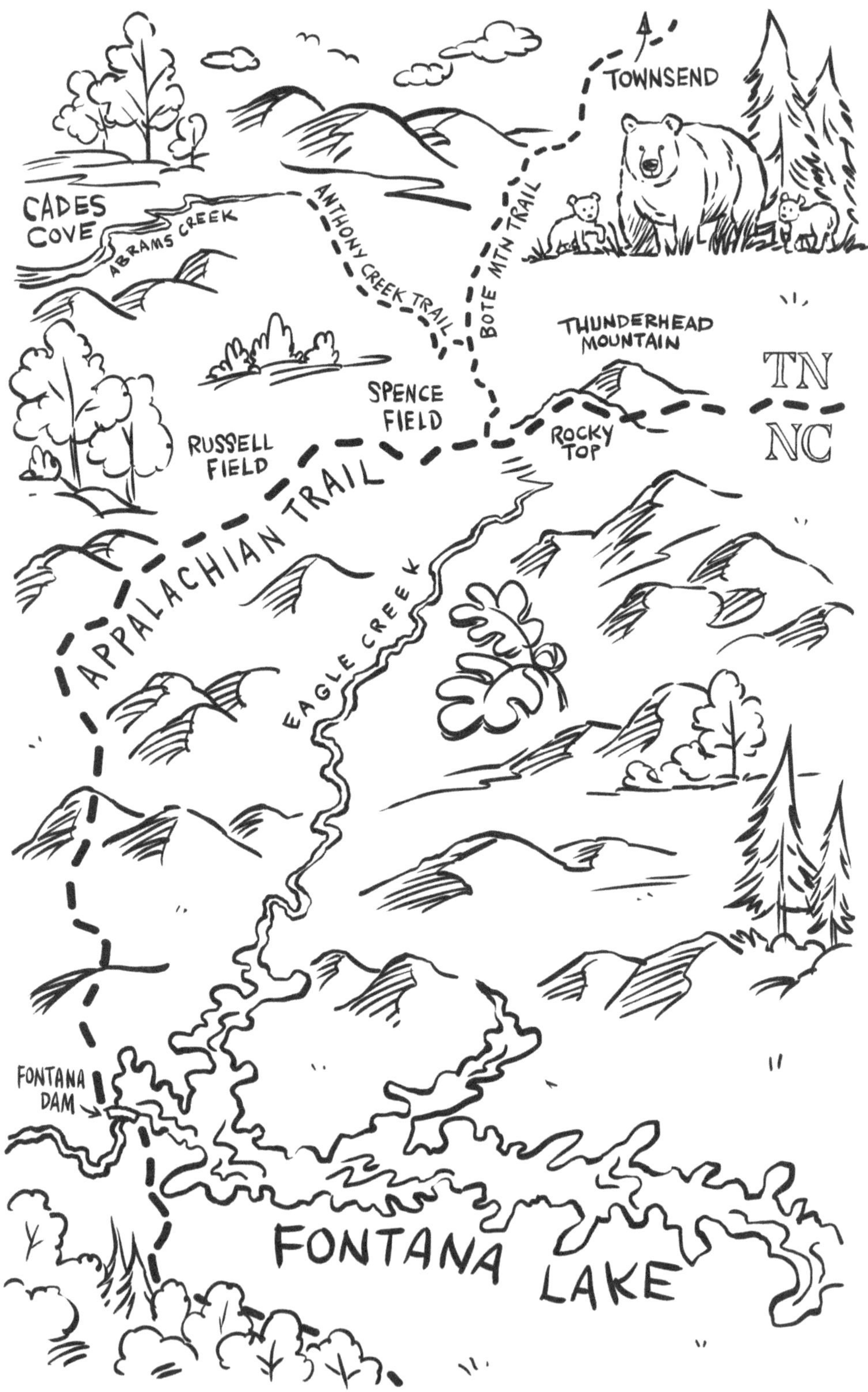
TOWNSEND
CADES COVE
ABRAMS CREEK
ANTHONY CREEK TRAIL
BOTE MTN TRAIL
THUNDERHEAD MOUNTAIN
TN
NC
SPENCE FIELD
RUSSELL FIELD
ROCKY TOP
APPALACHIAN TRAIL
EAGLE CREEK
FONTANA DAM
FONTANA LAKE

HEADWATERS, 1954

GOD GAVE ME GREAT PARENTS, and they had the common sense to let me be a boy (within certain parameters).

It was the spring of 1954, and I was twelve years old. I had followed my dad around the trout streams of the Smokies since I was seven and learned so much about fly fishing from him. I had been to summer camp in the park for several years and learned lessons about the mountains, hiking, and camping. And now my parents began to let me fish the Smokies on day trips on my own. Of course, I wasn't old enough to drive. But if I could prevail upon my mom or an older friend to get me there and back, I was free to fish all day on my own. The following year, they even let me do overnights on my own.

I began to wear out my older friends and family, begging to be dropped off in the Smokies for a later pickup. Mostly I fished the Little River from the Sinks to the three forks above Elkmont; the Middle Prong past Treemont, including Lynn Camp Prong; Sam's Creek (Thunderhead Prong); Abrams Creek in Cades Cove; and the West Prong of the Little Pigeon and its tributaries. My dad had said that I could fish wherever I wished provided I always showed up at the pickup point on time, which I always did. I didn't dare mess up a good thing.

But even with the newfound freedom, there was a limit as to how far I could go on day trips—and I wondered what lay beyond, up higher in the more remote valleys and the tallest of the mountains.

I fished day after glorious day that summer. Even as a boy, I became pretty good at Smoky Mountain creek fishing. Certainly not as good as I thought I was. But good enough for it to be a source of confidence.

Sometimes that's good for a boy.

But even beyond that newfound confidence, there was something else I was experiencing for the first time: a profound sense of freedom, something I had never fully comprehended or expected. I could go anywhere my legs could sustain me.

There were some great fly fishers in the park back then, and when I encountered one of them on the stream, I carefully watched the way they fished, and I learned from them. There were times that I saw things I couldn't fully understand, at least not at first, but it was magnificent to see. Of all the great fly fishermen I encountered in the Smokies, there was one that stood out from the rest.

It was a year earlier, in April 1953. My dad had taken me to his favorite river. It flowed along the southern edge of the park—below the old Calderwood powerhouse. This was before the construction of Chilhowee Dam, and long before the Tellico Dam.

It was the Little Tennessee River, also called the Little T. As the great river came around a bend, it flowed over a giant shoal that spanned the entire width. Like the crown jewel of the Smokies, it sparkled in the morning light—and to this day I have never seen a river more beautiful.

I was walking beneath the tall hemlocks when I spotted an older man. He was well out into the clear riffle, and something about him made me stop and watch. He had that tanned and weathered look of someone who had fly fished all his life. His hat appeared battered with age. It was cocked back on his head and the smoke from his pipe seemed to drift along in tandem with the current and the gentle breeze. And the way he cast the fly rod was like nothing I had ever seen or even dreamed of.

His casts were powerful and fluid in equal measure, and his fly line appeared to float lightly in the air and roll on and on, beyond the law of gravity—stretching out farther and farther as if guided by some force of nature I could not fathom and didn't know existed. Thousands of droplets exploded from his line and sparkled in the sunlight. Each stroke of his cast was smooth and purposeful and with the last forward stroke something magical and unexpected happened: a point at which he would stop the rod, release the line, and it would fly on and on as if by magic. Distances I had never thought possible. To me it seemed so natural, as if his cast was powered by the great river and the mountain breeze. I wondered how he did it and if he was the best fly caster in all the world. Surely, he was. He would have to be.

I was just a kid, watching in total fascination, and even at that young age, I realized that what I saw was magnificent, a perfect connection of man and nature, equal parts of one masterpiece. I stood there on the bank of that great river beneath the tallest of the trees and stared in disbelief, mesmerized by the artistry.

When he cast, it was like the rhythm of the seasons, something natural like the ebb and flow of some giant tide, and each stroke of his cast seemed effortless, in sync with the surging of the current and the sunlight dancing on the riffles.

Later that evening as Dad drove us on the winding curves of US 129 back to Knoxville, I thought about what I had witnessed, and I was convinced that no other person in all the world could cast like that. I hoped that one day I would be like him and that I would find the magic in my own cast. I knew that I must. I had to.

After that day on the Little T, I began to realize there were others like him on our southern rivers, and that fly fishing was a thing that southern highlanders took great pride in, something deep-seated, authentic, and intrinsic to the Smokies—a part of our mountain heritage.

It was one of those times in my life when I learned what was possible without learning how. But once I knew it existed and what it looked like, I worked to achieve it. This was before online videos and social media. Even the books on trout fishing were few. The magic for a young boy was in the experimentation and the discovery.

Sometimes I learned about fly fishing and the Smokies when I wasn't even near the mountains. On one such occasion, at the age of twelve, I was sweeping the floor as my afternoon job at the Tombras Group, which was my dad's advertising company in Knoxville. There was an artist working in our creative department named Lee Roberson. He was a native of Townsend at the edge of the park, and he loved to hunt, fish, and paint scenes of the Smokies. I was fascinated by his paintings and stories, and we became great friends.

It turned out that Lee had a profound depth of knowledge about fly fishing in the Smokies, having lived his entire life in the mountains near Townsend. He had fished every stream in the southwest end of the park as well as all the backcountry streams on the North Carolina side.

Every serious Smoky Mountain fly fisherman, even a twelve-year-old, at some point gets hooked on the possibility of better fishing in the remote headwaters. Since Lee worked at Tombras and was eight years older than me, I was confident my dad would approve me fishing the backcountry with him. Soon we were planning a ten-day hike across Spence Field and down into the very headwaters of Eagle Creek in North Carolina.

The day our trip began was a little on the cool side. It was early spring, maybe too early in the spring, but the weather report looked good.

We checked our gear before starting out. Like most mountain men in those days, Lee believed in walking fast and traveling light. So before we began the trek, he insisted that we divide our gear into two stacks that he referred to as a "can't do without" stack and a "wouldn't this be nice to have" stack. After Lee's informal inspection, we ditched my tent. We could snap our two ponchos together and make a tent if it looked like rain.

After culling a lot of my stuff, here is what I ended up with in my lightweight backpack: army blanket, poncho, one pair of jeans, one pair of socks, extra shirt, light wool sweater, lightweight camp shoes, canteen, one pound of bacon, three potatoes, three onions, three carrots, a bag of cornmeal, cooking oil, salt, bread, a few slices of country ham, hot chocolate mix, coffee, fork, small lightweight frying pan and cup, pocketknife, leader, two small fly boxes, small bottle of homemade dry fly flotant, small flashlight, several boxes of waterproof matches, small bottle of purification tablets, toothbrush, toothpaste, and a simple lightweight over-the-shoulder canvas creel.

Lee had the same except he also had a lightweight first-aid kit and a small ball of stout twine in case we had to rig a poncho tent.

We hiked in our trout-fishing wading shoes and wore the ultra-lights in camp at night.

Sure enough, Lee walked fast. We left Cades Cove in the late afternoon on the Anthony Creek trail and then took the Bote Mountain Trail trying to reach Spence Field on the very spine of the Smokies before dark. My twelve-year-old legs had to really stretch to keep up with Lee's longer gait.

Anyone who has hiked the Bote Mountain Trail knows that the top part is moderately steep, and with every switchback on the trail, you think you are almost to the top, but each time, the top remains enticingly and mysteriously somewhere else. It teases you and torments you after nearly six miles of uphill hiking.

At last we summited the field just as the sun was setting behind us in the west. Thunderhead Mountain and its lesser peak of Rocky Top loomed above us to the northeast, magnificent in rosy light, and the North Carolina mountains faded off more bluish to the south and east below. I had never been up here before; Dad and I had always been down in the valleys in the trout streams. I felt like I was on top of the world. And in that moment, it was a magnificent sight that only Lee and I could see—a world apart from the one we had come from.

We spread our ponchos with blankets on the grassy bald, cleared a fire ring, and cooked a dinner of country ham, thin-sliced potatoes, and bread. After dinner, we heated coffee on the open fire, topped it off lightly with just a drop or two of red-eye gravy from the frying pan, and sipped this unusual concoction in the fading light.

I rolled up in my poncho and blanket, and with a mound of grass as a pillow, I dreamed of the wild stream I had yet to see.

It was cold up there on the bald that early spring night, and a light breeze blew. But I was exhausted from the speed of the six-mile uphill hike, and I slept through most of it.

The sun came early on the high and grassy bald. We cooked bacon, potato slices, and coffee, put water on the fire, and soon enough we were packed and heading down the grassy bald into North Carolina.

Blue mist rose like smoke nearly to the mountaintops, endless below me as far as I could see, concealing the mystery of the stream we would fish in the long and distant valley winding its way into Lake Fontana. Lee called it the backcountry as there were no roads. They had all been covered up when TVA completed the dam and backed up the waters of Fontana Lake ten years prior.

The grass was lush and green. I walked close to some thicker bushes, and a covey of grouse exploded in every direction. I had seen grouse before but never scattering so high above two states and the valleys far below. It was beautiful beyond words—a reminder that we were entering one of the wildest parts of the Smokies without being that far from home. I liked the thought that Spence Field was so close to Knoxville. Already I was concocting plans to return here many times, though I've probably never set foot on Spence

Field more than seven times in my life—maybe fourteen, if you count the round trips to North Carolina. But just knowing that it was there was enough.

We entered the forest below with wet dew on all the leaves. It was quiet as we moved quickly down the steep trail. Soon I heard water. At first, it was just the light splashing and gurgling of tiny brooklets, and then the sound was louder—water gushing over rocks cut from the ages of time, but still too shallow for even brook trout.

Farther down the trail, two streams converged. And now there were little riffles and drop pools, dark under the rhododendron and sparkling where rays of light broke through the lush canopy. Most of the pools were shallow, but soon I began to see some with undercut rocks—perhaps even deep enough for small trout.

Then in one of the deeper spots we saw faint shadows gliding ghostlike from under an undercut rock—wild trout finning in the clear currents where the beams of morning light shone through the canopy. They were the natives, the original trout of the Great Smokies. All the mountain people that I knew back then, including Lee, called them "specks."

I rigged my fly rod with a little dry fly that we called a Thunderhead, and I wondered if it was named after the great mountain that loomed above Spence Field. I had tied it with wings of white calf tail along with a gray muskrat fur body and grizzly and brown rooster hackle wound in up front. The little fly rode high and bounced lightly through the fast water, and the brook trout seemed to strike it excitedly. We caught them in almost every pool, most of them only six or seven inches long.

The backs of these beautiful creatures were dark olive, flecked with yellow squiggly marks, and their sides were luminous and glistened with multitudes of round yellow spots of all sizes, some of them encircling tiny red dots. The bellies were white with a tint of orange, and the front of each fin had a flourish of white.

They had survived throughout these wild mountains for untold years, but now they lived only in the high places like this, where the water was cold and pure. There were far fewer brook trout down in the valleys—the places where the logging companies had been.

Lee fished above me, and I walked down the steep trail alone. Suddenly the sound of the creek was louder. I heard it through the trees before I could see it, a small brooklet cascading into the Eagle. Now the pools were a little deeper, and instead of six-inch and seven-inch brookies, some were slightly larger. It was steep, and I climbed boulders and little two-foot waterfalls one after another, and in nearly every pool there was yet another brookie.

Sometimes I crawled through dark, tangled masses of rhododendron, draped low over the stream and me, almost like a cave, and I used a "slingshot" style of cast beneath the tangle of limbs and leaves with limited success.

Finally I came to a high chute of water that surged through a crevice and fell eight or nine feet into the deepest pool I had encountered. I could clearly see several nice brookies holding in the current, and they were larger than the ones I had caught so far.

It was a short cast, and the little Thunderhead landed delicately in the clear current. The largest of them rushed the fly, and I hooked him. I measured it: ten inches, large for a brookie—the largest I had ever caught. Brook trout have a short life span; they only live four years in these small waters with limited food and little time in which to grow.

On each side of the waterfall were high boulders with no foothold, and beyond them on each side were cliffs too steep and sheer for climbing. It was one of nature's roadblocks. There seemed to be no good way around it.

I was confronted with two choices: The first was to backtrack down the stream, climb the steep slope up to the trail, and bypass the waterfall. The other was to try to climb the waterfall itself. The latter choice seemed the easiest. The waterfall was only about eight or nine feet. It seemed like a lot less effort than backtracking down the creek through the tangled mass of rhododendron and laurel.

The waterfall was narrow and the large boulders on each side stretched upward, clear to the top. The boulders were about three feet apart, and the water rushed between them. I wedged myself between the two boulders with my feet on boulder left looking upstream and my shoulders on boulder right. I began inching my way up. I was totally drenched by the waterfall but made it all the way up to the lip of the falls. Here the water had made the rocks slick. There was no more boulder behind me to wedge myself in, and nothing but small, slick, and loose rocks to hold on to as I tried climb over the lip of the falls. I hadn't anticipated that challenge.

First my hand grabbed a loose rock. Then I felt myself sliding from the lip, and I knew in a second that I was going to fall.

I fell the entire vertical distance of the waterfall, bouncing off the boulders on each side as I dropped. Instinctively, I tried to stay feet first to avoid a head injury when I hit the bottom. The thought flashed through my mind in a millisecond that this was going to hurt. And the thought was right. I landed full force on the rocks below on my right knee. I writhed in pain and agony for quite a while, now shivering from shock and drenched by the icy-cold water.

I tried to move my knee, but it wouldn't bend. I tried standing to see if the leg would support weight. It would, sort of, but only a very tiny bit. I was okay, but only if my weight was completely on my left leg. Looking around, I noticed my fly rod had somehow survived the fall and I hadn't landed on it. I picked it up.

What I was about to do was what I should have done when I was presented with the two choices: backtrack down the creek, climb down the steep boulders, and crawl through the rhododendron tunnel to reach the place where I had entered the creek and could climb up to the trail. Only now, due to my first decision, I was going to do it injured.

It was tough backtracking with a knee that wouldn't bend. It took an hour, but at last I got to where I had entered the stream. The trail was high above me but seemed reachable.

Now I started making better decisions. My knee was swelling fast and aggressively. I needed an ice wrap, which of course was impossible here in the Smoky Mountain backcountry. So I stopped for thirty minutes and soaked my knee in the cold rushing water of Eagle Creek. I heard a sound behind me, and Lee was sliding down the steep slope from the trail above to see what was up.

I quickly recounted the story and showed Lee the swollen knee. He said, "It's too far for you to make it back to Cade's Cove in Tennessee on that knee. Let's set up camp somewhere, give it two or three days, and see what happens."

That sounded good to me. I didn't want to ruin Lee's trip, and I wasn't at all anxious to leave this wonderous place now that I was finally here.

The slope was nearly vertical up to the trail, but with a pull from Lee every now and then, I made it.

Once on the trail, Lee said, "Stay here. I'm going downriver to see if I can find a good place to camp. I'll be back."

I sat down and rested with my back against the bank on the trail, listening to the cadence of the stream and feeling the solitude of this place.

Soon Lee was back and smiling. "Found a perfect spot. It's flat, off the trail, and close to the creek. Not more than four hundred yards downstream from here." Things were looking up.

With the help of a stout, forked stick that Lee carved into makeshift crutch, I limped into our new home on Eagle Creek. I took off my wet clothes and substituted them for the dry ones in my pack. The knee was starting to swell badly, so every hour or two I soaked it in the cold headwaters of Eagle Creek. I think that helped prevent the knee from getting worse.

Nevertheless, it was swollen to the size of a football. For the next two days, I stayed in camp with it elevated. We called this place the Eagle Creek Medical Center. I did nothing more than soak the knee in that cold creek water for the first day or so. After that, I watched clouds drift by above the tall trees, listened to the rushing water, and occasionally threw another stick on the fire to keep the hot chocolate or coffee warm.

There was a lot of time for me to think about the choice I had made here in the backcountry. At the very least, I absorbed the lesson about climbing waterfalls.

Each day the knee improved slightly. Lee fished in the mornings and evenings and came back with fresh trout and wild ramps. We sliced and diced the ramps—leaves, onion bulb and all—and fried them in the skillet with chunks of bacon, thin-sliced potatoes, and wild trout with a little salt. To tell you the truth, I had it made.

By the third day, the swelling had subsided some, and I was fishing but with a severe limp, which seemed to be improving. So we moved camp miles down the river, where the rainbows live, and fished most days from daylight to dark, even skipping lunch. For ten days, we never saw another human being. That's the way it was back then in the headwaters of the streams on the North Carolina side of the park.

As if by some miracle, when we finally walked back out of Eagle Creek and made the long climb up to Spence Field and down to Cades Cove in Tennessee, I was walking as if no injury had ever happened.

Lee continued to work at the Tombras Group for several years, and later our lives took us in different directions. Nevertheless, our friendship was lifelong and based on fly fishing and a love of the Great Smokies.

Though we were different in age, we shared a dream: it was the mountains, the hidden glens, clear waters running free over muted rocks, and the sparkle of the sunlight dancing off the riffles. The images of his dreams were in his paintings, and mine were in my memories.

Lee eventually built a rambling chink log home and art gallery on a wooded property with a split rail fence. It was in a mountain cove with a small brook and tucked back off Wears Valley Road near Townsend.

People who visited Townsend and were lucky enough to stumble upon his place were thrilled. He had a laid-back mountaineer way and was an authentic part of the Smokies.

Lee passed away in 2014. Some of his prints are still seen in the restaurants in Townsend. And there is one that haunts me, a fly fishing scene high in the headwaters. And if you look closely in that scene, you see the brook trout rising to inspect the dry fly. And I recognize that fly, and sometimes I think I know the pool, and I wonder if my old friend is there now, at that very place, watching over the headwaters where we fished.

BANDIT IN THE BACKCOUNTRY, 1955

A TWIG SNAPPED in the darkness.

It was an odd sound. A bit startling and unfriendly—not the kind of sound that just happens randomly in a forest every now and then. I listened for a while. There was nothing more, but I knew he was there—and that we were being watched.

I lay quietly in my sleeping bag. It was dark. "Daylight will come soon," I thought silently. "I'll stay awake just in case."

Soon enough, the first hint of gray dawn usurped the night sky. Even in the dim light, the woods remained dark and quiet. I hurried to get the fire going for coffee and to warm up. The sun comes slowly to the creek bottoms in the steep headwater valleys high in the Smokies.

I started the fire with a drop or two of my dry fly flotant. It was a concoction my dad had shown me how to make back in the 1940s, a mixture of lighter fluid and paraffin. It served double duty—an emergency and convenient fire-starter as well as something to douse my dry fly in occasionally to keep it floating on the surface while fishing.

Duane was soon awake, and the coffee was almost ready. We had been friends since the fourth grade at Bearden Grammar School. Though I was only fourteen and Duane was a few months older at fifteen, we had already fished the headwater streams in the Smokies together many times. These streams were home waters to us. And we hoped they would never change, even though at that young age, we sensed that someday more people would be here. For now, we savored the moments these places gave us.

I slipped into my fishing jeans, still wet from the day before, laced up my wading shoes, and sat by the fire, anxious for the warmth of the day. With visions of so many trout in this seldom-fished stream, I was too restless and antsy for a full breakfast. We would hike and fish close to fifteen hours a day this time of year, and yet each minute the day gave us seemed precious. We were in pursuit of something uncertain and ill-defined, but too important for us to tarry.

So a full breakfast was something we skipped that morning, opting instead for a quick piece of bacon on a slice of white bread. Then we were off down the river to see what it had in store.

When at last the sun came, it was wonderful, and the sky was clear. The day was full of promise and hope. We fished a mile or more, releasing all the trout. By noon we had fished our way back to our little camp, and with no substantial breakfast, we were famished. Duane started a fire in a stone ring we had built beside the stream, and once the thin sliced potatoes were browned, he was ready for trout. I tied on a size 18 Gremlin dry fly to my tippet and knelt by the stream near the fire, keeping a low profile so as not to spook the fish.

I flipped the little Gremlin into the gently swirling pool. A flash of silver and the rod bent briefly as I lifted the little seven-incher out of the water. With the rod, I swung the fish, still on the line, gently over to Duane. He removed the fly, cleaned the trout, applied the cornmeal and placed the fish in the hot frying pan. This entire sequence of events, from the hook set to the frying pan, took approximately forty-five seconds, maybe less.

By the time he had the fish in the frying pan, I had applied more flotant to my little dry fly and had recast to the pool. There was a splashy take. I lifted my rod tip and had a clone of the first rainbow that I was lifting with my rod over to Duane to clean and drop into the skillet. This sequence repeated itself three more times, until the skillet was full, with five trout. The total time elapsed was probably four minutes.

I climbed from the stream, and by the time I set my fly rod against the mossy outcrop, Duane had flipped the trout in the skillet, and they were browning on both sides. We sat by the green ferns on the stream bank, with our feet dangling in the clear current, and tasted one of the freshest trout lunches ever fried in the Smokies.

That afternoon we hiked miles and miles down this enchanted stream and released beautifully colored eleven-inch rainbows. To Duane and me this high mountain valley was wonderment, and the stream seemed alive—cold and refreshing against our legs, forever moving, like life itself. It gave us strength and we basked in the freshness of it, the same as all the other days on these backcountry streams. Together, we fished on and on for hours, entranced by the mystery of the forest and the moving water.

Somewhere in the riffles and eddies, flowing swift beneath the mountain laurel, there was a meaning to it all—there had to be. And at our young age, we fished this stream and all the headwater creeks, as if we were driven to find it—if not in the pool below the waterfall, then the next, or the next. We fished on and on, but we never found it. And yet we knew it was always there, swirling behind each boulder, stirring in the pebble depths of each dark pool, waiting in the shadows beneath the cool hemlock, farther up the river and around each bend, higher and higher into the Great Smokies where the blue mist hovered and the pure water sprang forth from the coolness of the earth.

We were young and strong and alive, and we fished mile after mile that day as if we were in a trance, with no idea that any other moment existed besides the one on this backcountry stream. It was on days like this, high in the headwaters, that I wondered if fly fishing was the reason God had put me here.

And then, at pitch-black dark, we hiked the long, thin path back to camp. Even later, after dinner, we told and retold tales from the day. It was our campfire ritual. It's how we learned from each other. To us, the world was new, and it seemed as if we would be here for a long time, and we sat by the fire and relived the day and others like it.

We talked late that night. And then, as the campfire burned low, I heard a sound in the forest, just like before, as if something unseen was out there and watching us. We were alert, and we listened. We saw nothing in the darkness of the forest and heard nothing more.

Nevertheless, as a precaution, we divided our gear between Duane's pack and mine, and kept the packs close by each other's sleeping bags that night as we slept. The rest of the night was uneventful, and the morning light came slowly, laden with a heavy mist.

When I awoke, I discovered that my pack was not where I had left it beside my bedroll on the ground. In fact, it was clear across our little makeshift campsite, and most of the food in my pack, but strangely not all of it, was missing.

There weren't that many bears in the park back then. We had not seen much reason for bear-safe ropes to hang the food packs. Today there are approximately four times the number of bears in the Great Smokies as there were that evening in 1955. Nevertheless, we were dealing with a thief. We were a long hike from civilization and needed our food supplies, and we couldn't risk losing the rest.

We planned to fish far from camp that day, and we didn't want to return and find that the remainder of our food had disappeared. So we carried it in our backpacks on the stream as we fished.

After another glorious day of fishing and a campfire dinner of wild strawberries, fried trout, potatoes, sliced onions, and skillet-toasted bread, we were once again ready to call it a day.

Duane was standing at the edge of the darkness, dimly illuminated by the last remaining glow of the campfire, and he said to me, "How are we going to keep that thief out of our food tonight?"

I replied confidently, "Don't worry, I sleep with one eye open. If he comes near me, I'll scare him off." Then I added, "Just to make sure, I'll use my pack as a pillow tonight. No way is that bear getting it."

I laid out my sleeping bag over the poncho on the ground, and with the pack containing our food as my pillow, I relaxed and looked at the stars. My mind wandered, and almost absent-mindedly, I thought about the bear that charged me at the Elkmont trash dump nearly six years before. I knew that these bears could live ten or fifteen years. Could it be the same bear? Perhaps, after all these years, our paths were crossing once again. It was an interesting thought, but soon I was asleep.

I slept soundly, and when I finally awoke, the sun was already rising. It was unusual for me to sleep late, but I felt refreshed and recharged. Then, as I stretched and moved my head, I realized something wasn't right.

My head was on the ground, not on the pack. I was sure I must have rolled off the pack during the night. But as I propped myself up in my bedroll and looked around, the realization set in: the pack was gone. It had vanished and was nowhere in sight within the camp.

My mind was racing. Was this even possible? Could this bear have sneaked in and stolen the pack right from under my head without waking me? The facts were indisputable; it was gone. But how could a two- to four-hundred-pound animal with foul-smelling breath be inches from my face, and me not know it? How could he pull on the pack so gently, so slowly and delicately, that he removed it from under my head without waking me? I was beginning to feel small.

Duane was already up, and he casually mentioned, "I see you slept with one eye open."

I caught the irony of his humor and managed a shallow laugh. I was feeling outwitted by an animal, and I had lost the precious food I had packed in all the way from Tennessee.

We searched the entire campsite. No pack. We searched up and down the stream for hundreds of yards, and then even farther. No pack.

Finally, we climbed the steep mountainside through the trees behind camp. About a quarter of the way up, we hit a thin trail of cornmeal.

We followed it through the steep foliage for a hundred yards to a ripped-open container of shortening. The trail was getting hot. Finally, the thin trace of cornmeal led us to a thicket of rhododendron and laurel. There, in a tunnel of shade, on the bare ground, we found my pack, ripped apart with the cornmeal, potatoes, carrots, onions, bacon and country ham all gone. He'd even eaten the salt. To add insult to injury, he had ripped up my spare T-shirt, angry perhaps that there was not more food for him.

Thanks to the bear, we were now nearly out of supplies. Our shortening for frying the fish and pretty much every other food item we had packed in was gone. All that remained were the few items in Duane's pack—a small bag of flour and coffee.

We fished hard that day, far from camp and up high in the headwaters. With no breakfast or lunch, by nightfall I was famished. We were out of food except for the small brook trout we roasted on sticks over the open fire. I had never cooked them that way, and they were tasty, but it left me hungry and wanting more.

As the campfire dimmed, I lay in my bedroll dreaming of cornbread, fried onions, and country ham. It didn't seem to help, but soon enough I was nearly asleep. The sky was pitch-black, the breeze had died, and in the darkness of the woods, I heard a sound. It wasn't much. Just a faint rustle in the thicket. And I knew he was there. Not far away from me.

Watching. Waiting. Taunting me. Mocking me. And I imagined him out there in the darkness, fat and happy from the food, and amused at my ineptitude. I knew he wouldn't come back into our little camp. He already had what he wanted.

I awoke at midnight when the rain came. It hit me lightly at first, and then grew heavier, raining hard the entire night. At first light we were cold and wet. The river had risen a foot or more and was off-color. There would be no fishing, which meant the end of our natural food source should we stay and try to exist with no supplies. We knew it was time to leave. Reluctantly, with no breakfast, and defeated by the weather, the rising water, and a recalcitrant bear, we packed up what little we had and left in the gray dawn.

As the storm clouds faded and the sun peeked over the Carolinas, two lean and ragged Smoky Mountain trout fishermen hiked across the highest ridge in the Smokies. They were headed back to Tennessee, cold and wet yet grateful to have been in this wild place. And though we were only boys, we felt like men, and from a distance perhaps we had that look. The appearance of men who knew the backcountry, who had followed the stream to its source and had found the truth. Indeed we had found it, and the truth was that we were just two teenage boys, cold, wet, tired, and hungry and looking forward to burgers, fries, and milkshakes.

N
W
E
S

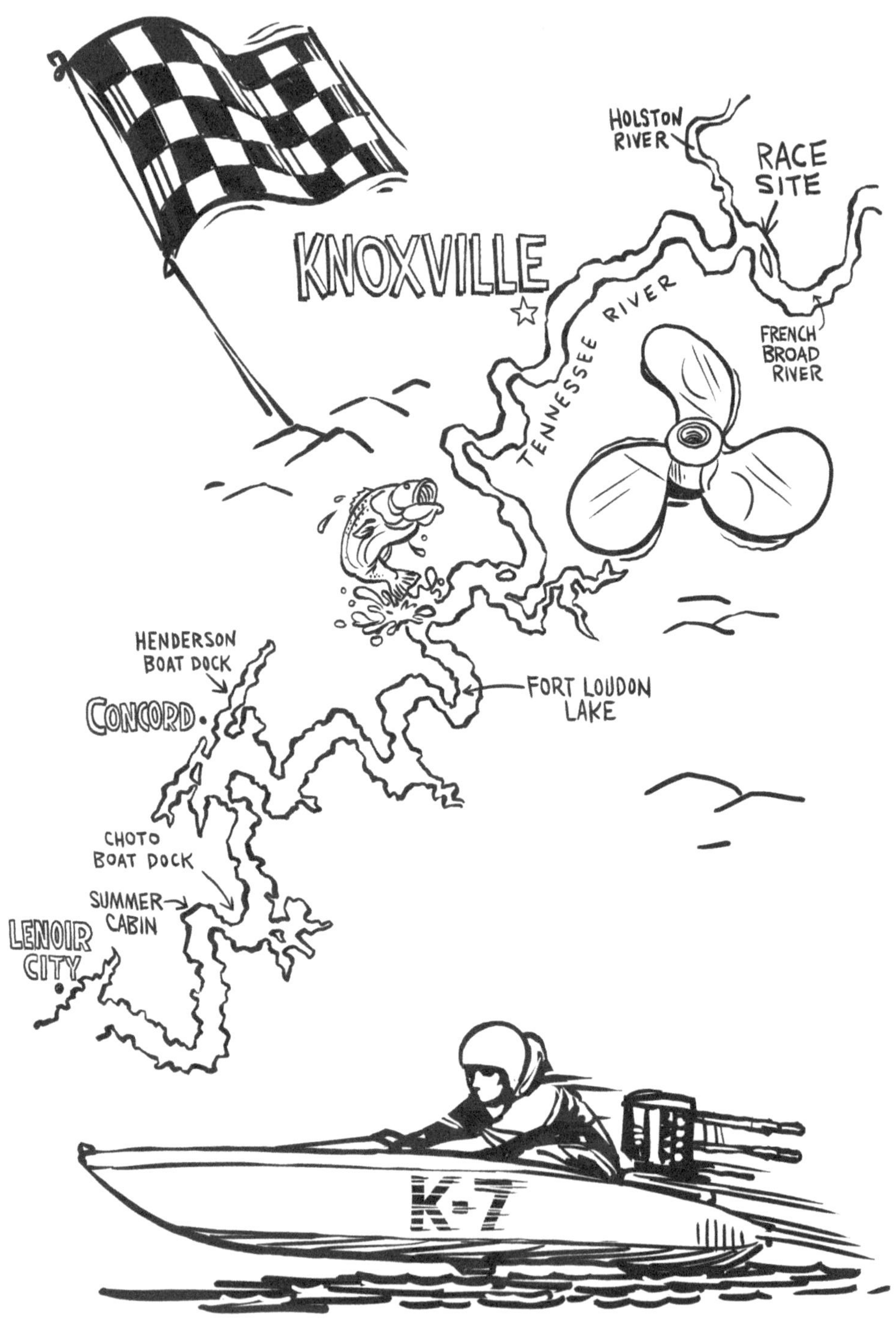
HOLSTON RIVER
RACE SITE
KNOXVILLE
TENNESSEE RIVER
FRENCH BROAD RIVER
HENDERSON BOAT DOCK
CONCORD
FORT LOUDON LAKE
CHOTO BOAT DOCK
SUMMER CABIN
LENOIR CITY
K-7

NEED FOR SPEED, 1956

THERE WAS A TIME in my life when I wanted to go very fast on the water. It was not a need I was born with; nothing about it was innate or inevitable, although I did come by it honestly. But by the age of fourteen, I was consumed by it.

It had all started innocently enough, three years earlier, with a gift from my dad: his 2.5 horsepower Johnson Seahorse outboard engine. It was an old one, very old, with a rope starter on an exposed flywheel. I was only eleven years old the day he gave it to me. It was September 1951, and we had spent the weekend at our little cabin on Fort Loudon Lake near Lenoir City, Tennessee.

The little cabin was nestled in a quiet cove in the tall pines. It was in a cove off the main channel two bends in the lake downstream from Choto Boat Dock (later called Choto Marina). The little lake cottage is where I grew up in the summers and on weekends when I was not at camp or in school.

Dad and I stood together on the old wooden dock, and he said, "Son, it's time you had your own outboard engine. I want you to have the little Seahorse. Take care of it." He had just purchased a brand-new 5 horsepower Johnson outboard engine for his square-stern Old Town canoe, and he no longer needed the little Seahorse.

I hadn't expected a gift, and at the age of eleven, I was overwhelmed. To me, the little Johnson Seahorse meant freedom. It meant possibilities. I could put it on the spare canoe and venture out on my own and go anywhere on the lake that I wanted, or so I thought.

There were very few boats on our East Tennessee lakes back then. Some days you'd never even see one. It was all pre-internet, pre-social media, pre-bass boats and lake homes. Most of the lakes around Knoxville were thirty miles long or more. On Loudon Lake, there were hundreds of miles of shoreline and coves, nearly all of it nestled in rolling wooded hills and farmland. It was a wondrous place for a young boy with a boat, and I longed to explore all of it.

I soon discovered that exploring a lake of that size was easier said than done. For one thing, the little Johnson Seahorse was cantankerous and unreliable. It was risky to venture very far from our dock in case I had to paddle home. Even more problematic was the speed—or the lack thereof. It was agonizingly slow. It was ironic that the little 2.5 horsepower engine that enabled me to explore, and teased me with my newfound freedom, was so slow that it prevented me from my dream of exploration. In short, I now had the freedom to roam the entire lake, but not the speed with which to do it—and that frustrated me greatly.

That was when I first felt the need for speed on the water.

One day my dad and I went to Henderson Boat Dock for an afternoon of crappie fishing. It was well back in Concord Cove on the north side, where Fox Road Marina is now. That was the day my dream of going fast on the water began to materialize.

Old man Henderson owned the ramshackle farmhouse and dock. He reminded me of the western movie actor Walter Brennan. He made his money by renting boats to fishermen, and every angler, upon arriving at his little dock, always had the same question: "How's the fishing?"

Old man Henderson had a pet largemouth bass that lived under the dock. When people asked him how the fishing was, he would casually tap his foot on the dock to alert the bass. Then he would take a couple of shiner minnows from his bait tank, toss them in the water, and that huge eight-pound largemouth bass would rush from underneath the dock and engulf them—leaving the visitors gasping and wanting to immediately rent one of his boats and go fishing.

But even more interesting to me was his lightweight marine plywood runabout with a 16 horsepower Scott-Atwater outboard engine. It was brand new, and it would fly, or so I thought, maybe twenty miles an hour. I coveted that boat. He offered to let me drive it. I was thrilled. I was hooked. I was a teenager driving one of the faster fishing boats on the lake. Who could have foreseen the speed of today's bass boats?

The following year, my family bought a little fourteen-foot runabout with a 25 horsepower Buccaneer engine. One of my good friends from Bearden School was a boy named Jim Farnham. His family had purchased a boat just like ours, and Jim and I learned how to water-ski. I even got a job up at Concord giving water-ski lessons.

Naturally there was a rivalry as to who could get their little boat to go faster. Pretty soon Jim and I had them running 30 mph, and we thought that was almost like flying. It was about as fast as pleasure boats went at that day and time, and it was all very exciting. I yearned for even more speed.

Later that same year, I discovered Gene Whitten. He owned Choto Boat Dock and had raced hydroplanes when he was younger. Gene had a 40 horsepower Mercury four-cylinder engine on a lightweight AristoCraft plywood boat, and it would fly. It was the fastest boat down at our end of Fort Loudon Lake.

Gene took me under his wing and taught me how to tune outboard engines, and especially the propellers, to give the boat extra lift, RPMs, and speed. He taught me the ins and outs of boat racing, the different race divisions and classes and tips for race driving.

In early 1956, he helped me locate a six-cylinder Mercury engine, which we put on a much faster boat, a fourteen-foot Crosby, in a special model built for speed. Surprisingly, I found myself driving one of the faster outboard pleasure boats in the entire country. Not many people knew that. But it was significant because it was the year that the National Outboard Association began a new division for boat racing. It was the equivalent of NASCAR in automobile racing.

The inaugural race in that new division was set for Three Rivers Boat Dock, where the Holston and French Broad rivers came together to form the Tennessee just above Knoxville. The National Outboard Association published all kinds of race regulations for the new division, including engine and boat modifications, and page after page of race driving rules. However, in all their rules and regulations, the one thing missing was any kind of age restriction for the race drivers. I took note of that. It was the first thing that I looked for. I was only fourteen.

With some financial help from Gene Whitten, I purchased a racing jacket and helmet and sent in my entry form for the race. The only problem was, I was too young to drive on Tennessee highways. I had no driver's license, so getting my boat to the race by trailering was out of the question.

I solved that problem by getting an early start the day of the race and running my boat there by water. It was about thirty miles from our lake cabin to the race site at Three Rivers.

I remember distinctly that my sister, Nancy, Dick Powell, a good friend of mine from Bearden School, and Gene Whitten were there by race time to cheer me on. But at the age of fourteen, I didn't know many others at the race. In fact, I don't recall that I knew anyone else among the large crowd of spectators and race crews.

I arrived in time for the 10 a.m. drivers' meeting. About a hundred drivers and boats showed up for the race. There were five classes. Boats were assigned to race classes based on the engine cubic-inch size, boat weights, lengths, and other factors. They put me in the fastest group, the Unlimited Class.

I looked around the large room at the drivers' meeting. The other drivers were in their twenties and thirties, with a few in their early forties.

Their average was at least twice my own. The driver that seemed to command the most attention was Jimmy Rogers. He worked at Fox Marine on Broadway and was sponsored by Fox with the backing, expertise, and technology from Mercury. He had the factory training, tools, and resources for his boat to run the fastest it possibly could. I overheard some of the other drivers say he had by far the fastest boat at the race and in the entire United States.

Fortunately, the race officials never asked my age. I was barely even noticed, which was fine with me. I stayed near the back of the room at the drivers' meeting and didn't talk to anyone, as I was concerned that if they realized my age, they might have me disqualified.

A huge black-and-white ten-foot clock was placed at the end of one of the docks at the starting line, along with a medium-sized cannon. For each race, the officials fired a five-minute warning with the cannon. This loud boom and the smoke alerted the drivers to get their boats out of the pit area so they could mill around in the river upstream and far back from the starting line.

Then the race officials fired a loud one-minute warning shot on the cannon. At this time, it was legal to go back a mile or more from the starting line and run full speed toward it. Meanwhile, the second hand on the big clock would start at sixty seconds and work its way to zero. If a driver crossed the starting line before the second hand got to zero, he was disqualified. If a driver crossed the starting line when the second hand got to zero, or after that, he had a legal start. The perfect idea was to hit the starting line full speed right at zero. It took some practice. I had none.

The unlimited class had sixteen entries. The roar of the engines was overwhelming as we raced side by side toward the starting line. I couldn't believe this was happening. I was in a real boat race. My adrenaline was pumping like never before. It was a wild start with everyone jockeying for position, and I purposely slowed very slightly at the starting line to be sure I didn't jump the gun and get disqualified. That slight loss of speed put me back in the middle of the pack. Nevertheless, all sixteen boats hit the first turn at almost the same time, and it seemed to me there was room for only one of us.

It was like hitting a solid wall of water—like fire trucks with high-velocity water hoses blasting your head. The water hit me so hard it nearly ripped off my helmet. Visibility in the turn came in alternating bursts of light and blinding blasts of water from the boats in front of me. In the seconds when I could not see, I tried to judge the distance to the boat in front of me and others around me by the sound of their open-exhaust engines and the intensity of the rooster tails and spray tearing at my face. The harder it hit me, the closer I was to the boats ahead.

I passed four boats in the first turn and then my boat went airborne from the rough water and the power of the rooster tails of the boats ahead. As my boat went high in the air, another driver's boat was underneath me, and somehow, I missed him to the left at the last millisecond, even though I was sure I was coming down right on top of him.

Coming out of the first turn there were four boats in front of me, and I was running fifth. I tried to find smooth water where possible rather than run in their rooster tails, that way my prop could get a better bite and increase my speed. I found a smooth line and felt the immense acceleration and raw power as my boat lifted from the water, and the only thing touching the surface was an inch or so of the bottom rear of the boat and the propeller. It felt like I was flying, and one by one, I passed three of the four boats on the back stretch.

The lead boat was still way out front, and I could see it was Jimmy Rogers. We hit the far turn with Jimmy already nearly out of the turn by the time I rounded the first buoy.

I followed him down the front stretch and into the turn just past the starting line. This time he was only halfway through the turn when I roared past the first buoy. I was gaining.

A race is three one-mile laps, and I was in a strong second place during lap two and gaining on Rogers. I passed him near the end of the back stretch and rode out the rest of the race in first place.

I was mobbed by people and officials in the winners' circle and pit area who didn't know me and hadn't spoken to me prior to the race. Some of them were race boat owners and they were trying to persuade me to drive for them or even to move up into the professional hydroplane division. People even asked for my autograph.

Timidly I signed their race-day programs. They had no clue I was not yet fifteen. I think they all assumed I was older and knew what I was doing.

It was too much for my mind to take in. On the long thirty-mile ride back down the lake to our cabin near the dam at Lenoir City, I was one happy race driver. I had my trophy and the cash prize and felt I had surely become a man. Or at least that's what I imagined.

When I arrived at the little cabin on the lake that evening, I couldn't wait to tell my parents what I had accomplished that day. I imagined their excited reaction and how proud they would be of me. Their response was underwhelming and not what I expected. They hadn't a clue where I had been in my boat all that time or what I had been doing. At first it was disbelief, but my claims were later substantiated by my sister, Nancy. Then their reaction changed to relief that I hadn't been killed or injured and finally it changed to something else—guarded and reluctant acceptance without encouragement. And that was how they dealt with it from then on.

For years afterward I continued racing boats on the weekends on the National Outboard Association circuit. There were races that I won and a whole lot more that I lost and a few where I flipped the boat or crashed and ended up in a hospital or the medical tent. But to this day, the race[3] I remember the most was the first one—the race I won before I even had a driver's license.

---

3. The year after the race in Knoxville, the author moved up to the professional hydroplane division where the racing engines had fewer restrictions and burned exotic fuels—mixtures of benzol, methanol, nitromethane, and propylene oxide. Some of those boats were fast for that day and time, over one hundred miles an hour.

N
W
E
S

ROYAL COACHMAN
ABRAMS FALLS
CADES COVE
ABRAMS CREEK COVE
TO MARYVILLE/ KNOXVILLE
129
US 129 BRIDGE
BUSHWACKING TO PANTHER CREEK
PANTHER CREEK
PARSONS CREEK RD
WILDERNESS AREA
NO DESIGNATED CAMPSITES
RAINBOW TROUT
129
CHILHOWEE LAKE
TO FONTANA
129

UNTOUCHED WATER, 1959

LONG BEFORE THE ADVENT of the internet and Google Earth, fly fishers in and around the Smokies purchased enormous collections of topographical maps. Their goal was to find that one small, secluded creek, so far from roads, trails, or campgrounds that it was overlooked, or somehow so inaccessible that it was unfished and therefore harbored trout untouched by man.

At the age of seventeen, I was one of them. I spent hours at nights and on the weekends studying these maps, looking at the skinny blue lines that were the streams, hoping to find that hidden jewel. Unfortunately for me, there were many others with the same dream, and we all had the same maps, all searching with the same purpose.

Mostly, it was a snipe hunt—a dream of something that didn't exist. But occasionally, I heard rumors that someone had found something—second and thirdhand accounts of someone who knew someone who knew someone else who had made a discovery. A secret and remote rivulet high in the mountains. A small brooklet that appeared too shallow for fish. No trail or path, but according to the rumor, there were waterfalls farther up this tiny brooklet where the pools were deeper, and the trout were there—trout so wild they had never seen an artificial fly. The rumors were rarely true, and most of these fly fishers never found what they were searching for.

But I did—although it was only once.

The waters of the lake were backed up in 1957 when Alcoa completed the Chilhowee Dam and flooded a large section of the Little Tennessee River

(Little T) all the way up to the powerhouse at Calderwood. As the waters backed up into Abrams Creek Cove, the park service put a net under the new US 129 Highway bridge to keep boat traffic out of the park.

What was unknown to many, and certainly not obvious to anyone passing by, was that there was another creek in addition to Abrams Creek that flowed into the main cove out of sight to all. It was known as Panther Creek. It wasn't a great trout stream prior to the lake because people had access. But with the new lake covering all the roads as well as the boat barrier placed by the park service, all access had been quietly eliminated.

It took only a few years without fishing pressure for the trout population in Panther Creek to increase dramatically. The resurgence was aided by rainbows from the new lake that swam up Panther Creek to spawn.

There were no trails into Panther Creek. The barrier net beneath the bridge on the lake meant there were only two ways to get there, and both were bad. The first was to drive farther up the highway, park the car, and bushwhack up and over a steep mountain and then down the other side to the lake. You came out short of Panther Creek and then had to wade chest-deep water in the lake for a good while to finally arrive at the mouth of Panther Creek. Very few people were willing to go to that much effort or could even find the route.

The other option was to drive the narrow Parsons Creek gravel road from Cades Cove toward US 129. It was a rough winding road, and it was one-way, so if you parked on it to bushwhack into the trailless headwaters of Panther Creek, you couldn't return by road the way you had come. It complicated things. In addition, the entire top end of the creek, as it flowed under Parsons Creek Road and downstream for the first several miles, was choked with blowdowns—dead trees that fell completely across the river. And each of these dead trees was infested with copperheads. There was no trail, and both sides of the creek were so steep, you had to crawl though those trees to work your way down the stream. I only went that way once. It was many years after the time of this story. I was with my first son, Charlie, my good friend Tim Keller, and his son Vince. I have fished a lot of places in the park, and that was the snakiest place I ever saw. I was glad we all got out of there without being bitten.

Once the lake was backed up by the new Chilhowee Dam, it was only a few years until Panther Creek became a hidden gem. It was rarely fished due to the difficult access and the fact that most people didn't know it was there.

The few that did know about it were, for the most part, unwilling to make the arduous effort to access it. In fact, in all the years when the cove was closed to boating, I only saw another fisherman once. As a result, the quality of the wilderness experience was truly exceptional.

The faint little path up from the mouth of the creek was not maintained. It was hard to find and follow in places. There were no designated campsites. The park had classified it a wilderness area.

I was a freshman at the University of Tennessee in 1959, three years after they began filling the lake. I had seen the barrier net at the bridge. I knew Panther Creek was there from my days before the lake. I had also seen Panther Creek on my topo map and was reminded that it was now inaccessible and out of sight.

Early one morning before daylight, I left Knoxville to check it out.

I drove through Maryville and up US 129 past the Little Tennessee River, which was still flowing free downstream from the new Chilhowee Dam. I passed the island across from Hoss Holt's house, where he later built his little store. Mist rose from the river, which was running minimum flow. I had already fished this great river many times on the shoal behind the island, the old trestle shoal, and the one downstream from there. It was hard to drive past the Little T that morning. The fishing was so good there, and not a soul was out on the river. Little did I know that I would only have nineteen more years on that great river, perhaps the best trout river in the eastern United States, until it too was dammed and gone.

But on this morning, I drove on by. I had another goal, and it was in the park.

I had studied the topo map, and it seemed that the second gap on the mountain past the Abrams Creek bridge would be the best route. I parked my car off the road. There was no trail. It was pure bushwhacking as I climbed the steep mountainside. It was a little farther than I expected to the top of the ridge, but at last I reached the crest. In front of me it was just as steep, dropping off abruptly through the trees to the lake unseen below.

Soon enough I arrived at the lake. The forest-clad banks were too steep for bushwhacking. Carefully I waded out into the water, aiming to the right toward where Panther Creek should be. The water was too deep for easy wading. It was to my chest. But the farther I went it gradually began to shallow out, until finally it was below my waist. I rounded a bend in the lake, and there in front of me was the mouth of Panther Creek with its current feeding out into the lake.

I rigged my fly rod with a tapered 5X leader and a size 18 dry fly called a Female Adams. Moving slowly and carefully to make no waves, I eased through the water of the lake toward the current feeding into it. There was plenty of room for my back-cast as I stood well out in the lake and sent a long cast about halfway up the lengthy run feeding into the still water. To my total amazement, on that first cast, a rainbow nearly thirteen inches long rose and delicately sipped the little dry fly.

The entire rest of the day was like that. I never fished more than a quarter mile. There were fish of that size or close to it in every pool, and usually several.

I kept two of the smaller trout for lunch and found a large flat boulder midstream that looked like a picturesque place to stop. I set up my Optimus cookstove in this newfound Shangri-La and fried the trout with sliced onions in cornmeal, salt, and oil. I topped it off with pan-seared white bread and a MoonPie for dessert.

While savoring my trout, I sat on the edge of the boulder with my feet dangling in the cool, clear current. After a few minutes, I saw two large trout enter the tailout of the pool and slowly swim past me upstream and out of sight. Were these trout moving up from the lake? Was it a spawning run? It was well into the summer, and rainbows almost always spawn in the spring, though there are exceptions. One thing was clear: these fish were moving up from Chilhowee Lake.

Later that afternoon I came to the first set of falls, more like a slide dropping about six or seven feet. I released three nice rainbows below the falls and then climbed up to the pools above. Here there were fish after fish, native rainbows, slenderer and brighter in color than the lake-run fish below, but every bit as long in length.

The sun was low, and I reeled in my line knowing that I had at last found that one trout stream in the Smokies I had searched and hoped for, hidden from civilization yet so close to it. Over the next fifteen years I mostly fished this stream alone, but sometimes I shared it with my dad or Duane Dunlap and occasionally with Graham Hunter, a good friend of mine and longtime fishing buddy from Knoxville.

Years later, the park service removed the barrier net below the bridge and opened Abrams Creek Cove to boating. Immediately the fish population and the quality of the experience in Panther Creek declined dramatically. I accessed it by boat a few times, but it was no longer the same stream. Each time, I had to hike farther and farther upstream to find good fishing and to avoid the growing crowd of people fishing the lower end.

After fifteen years of the best fishing that I ever saw in the park, my untouched water was lost forever, surrendered to the legions from the land of shopping malls, condominiums, and asphalt parking lots. Once the cove was open to boating, they came slowly and then in multitudes, and I never returned.

The riffles had lost their sparkle, and the magic was gone.

NEVER QUIT, 1955 TO 1964

AS I GREW OLDER, my parents wanted me to have a good education and must have felt I needed more discipline in my life. When I was fourteen, they sent me to a military academy and enrolled me as a freshman at McCallie, in Chattanooga, Tennessee. I guess they decided that if they got me away from rivers, lakes, boats and fly rods for a while, the more likely it might be that I would someday turn into an actual productive member of society.

At the age of fourteen I had no idea who I was or what I would become. I had this vague notion that my life would include fly fishing and boat racing, but beyond that, I was clueless as to whether there was any significance to my life, school, or the challenges that awaited me.

I didn't like being away from home at a military academy, at least for the first year or two, until I got better at sports. I missed my friends in Knoxville and fly fishing with my dad. I hated wearing a uniform, and whenever we had a free pass to go into downtown Chattanooga, I felt dorky compared to the kids from the public high schools.

Academic standards were very high at McCallie and required a lot of studying. It was something I struggled with at first because I was missing the mountains and my friends, and the excitement of boat racing. By the end of my junior year at McCallie, I told one of my teachers, Major Arthur Burns, that I would not be coming back for my senior year.

He asked me why. I told him that it was because I missed fly fishing and my friends in Knoxville. He looked at me for a long while and said, "Son, you've put three years of your life into getting a degree from McCallie, and you only have one more to go. But that's not what's important." He paused again and let that sink in. Then he looked me in the eye and said, "What's really important is what kind of person you want to be known as for the rest of your life."

It caught me off guard. It wasn't what I expected him to say, and I looked at him quizzically because I had no idea what he meant.

He said, "Do you want to be known as a quitter or a finisher?"

"If you quit now," he went on to say, "the next time something hard in your life happens, it will be easier for you to quit. Each time after that when you face adversity, quitting will be even easier. Soon, quitting will be a habit, and you'll be a quitter in life, and you'll be known as a quitter."

He said to me, "Think about how you want to be known and thought of in your life. Do you want to be known as a quitter or a finisher?"

Then he gave me the best advice I have ever received. He said, "The best way to never quit in life is to never start quitting."

Beyond that, he didn't pressure me. Instead, he simply asked me to go home to Knoxville over the weekend and think about it—and to come back Monday morning with my answer.

I thought about it long and hard over the weekend, and all the while, I knew what my answer would be. Monday morning, I was back in his office with my decision. And I told him, "I'm staying at McCallie for my senior year."

It was one of the best decisions I ever made.

There were a lot of teachers who helped me in many ways through the years, but none of them made a greater positive impact on my life than Major Burns.

In retrospect, McCallie taught me a lot of things, like honor and discipline. And for me, it was a slow and painful process. But the most important lesson of all was the one I learned from Major Burns, and the choice to persevere became habitual—a part of me and what I would become.

When I graduated from McCallie in 1959 and arrived at the University of Tennessee, I found there were other choices to make. One of those choices was whether to go through ROTC (Reserve Officers Training Corps). By going through ROTC, you could enter the military as an officer with a requirement to serve two years rather than risk being drafted and go in as an enlisted man.

I chose ROTC, as did a good many of my UT classmates, who seemed to excel at it. In fact, my good friend Sam Furrow, who had come to the University of Tennessee to play basketball, became the head student officer in ROTC. He then went into the Army, jumped from airplanes, and later was inducted into the University of Tennessee ROTC Hall of Fame.

I graduated from the University of Tennessee in 1964 and chose infantry as my first-choice branch in the Army, reporting to Infantry Officer Basic Training at Fort Benning, Georgia. It was three months of training on how to lead a platoon in conventional combat where there is a front line. There were three hundred officer trainees in my class.

It was all very exciting and challenging. My focus was to become a good officer and soldier. However, my goal in the Army was something more: a dream that had been simmering within me since those summer reading books in grammar school—the ones about World War II, like the Ranger battalion that scaled the cliffs at Normandy and the paratroopers who jumped into the Netherlands in the Battle of the Rhine. Quite simply, and for no other reason I've ever identified other than those summer reading books, I wanted to be an airborne Ranger. It was the reason I chose infantry as the branch of the army I wanted to serve in.

I quickly discovered that it helped to have the endorsement of staff at Infantry Officer Basic Training when I applied to Airborne and Ranger Schools. The staff placed a lot of emphasis on how you did on the PT course (physical training), morning runs, forced marches, and the final mile race in combat boots to determine who they recommended for Airborne and Ranger Schools. They only recommended the toughest soldiers. The rules weren't written down anywhere, but that was how it seemed to work at the time.

I had maxed all the physical training courses, scoring the highest possible on each component. For me the forced marches at night and the morning runs were easy, and I had scored at or near the top in patrolling. I was particularly good at map-reading and compass work (this was pre-GPS). I knew my way around in the woods due to all the time I had spent in the backcountry of the Smokies. There was only one thing left: I needed to make a big statement in the final one-mile race. I knew the entire staff would be out there on race day to observe.

One of the best officer candidates in our class was a fellow named Jim. He was about six-foot-two, extremely strong, very athletic, and had long legs. Everyone considered him a shoo-in to win the mile run. He and I were on friendly terms. We were approximately equal on all the PT courses, tests, and platoon-leading abilities. I suspected that I would get the full attention of the staff if I could win this race.

Knowing Jim and how competitive and confident he was, I figured he would jump out into the lead and hold it the entire race. Sure enough, that's the way it started.

My strategy was to shadow him and never let him get more than about fifteen feet in front of me. I didn't want to press him and make him speed up, but I also wanted to be close to him near the end of the race. I knew he had never seen me sprint and had no idea I could sprint fast near the end of a long run. My plan was to pass him coming out of the last turn and outsprint him to the finish line.

Which is exactly what happened. Somehow, I managed to beat the entire class of three hundred officer candidates.

The next day, the staff assured me they had been in touch with the Airborne and Ranger Schools and that I would be accepted as soon as I was assigned to a unit and had the unit commander's clearance. They said I would first go to Airborne School and then to Ranger School, in that order. That was fine with me, because I could use airborne training to get in even better shape prior to Ranger School, where being in excellent physical condition was even more important.

I was assigned to the Second Infantry Division and got the clearance from my unit commander. I was accepted to Airborne School and the Ranger School in that order. Both were at Fort Benning in Columbus, Georgia, where I was already stationed.

PART THREE

Storm Clouds Rising

WELCOME TO
FORT BENNING
U.S. ARMY
MILITARY RESERVATION
AIRBORNE

THE SILENCE BEFORE THE STORM, 1964

THE US ARMY Airborne School was the very heartbeat of Fort Benning, and it was easy to find. Like giant guard dogs, the three 250-foot jump towers stood tall above the base and could be seen for miles—a permanent and continual reminder to soldiers who aspired to be paratroopers about the challenge they would face.

I had heard stories about how demanding Jump School would be. I had also heard that it wasn't nearly as demanding as Ranger training. Nevertheless, there would always be the nagging fear of heights and the knowledge that sooner or later you would find yourself standing in the wide-open door of a military aircraft flying high above the state of Georgia.

At twenty-two, I felt more than ready. But I was in for a rude awakening. The military discipline in Airborne School far exceeded that in Infantry Officer Basic Training. The expectations were higher, and it was a jolting reminder that paratroopers were head and shoulders above regular "straight leg" infantry.

I quickly discovered that in Jump School, the training and hazing went on and on throughout the day, from the early morning formation and ruthless inspection to the runs and physical training, transitioning right into jump training, which lasted until just before dark. By night we were exhausted, and though we were free from the training, we were too worn out to do much more than sleep.

There was a lot of running. You ran everywhere in Airborne School. It was called the "airborne shuffle"—a slow run around the speed of an eight-minute mile.

The cadre were sergeants, crusty and tough, mostly combat veterans from the Korean War. They were all in great physical condition and wore black hats that set them apart from most everyone at Fort Benning. They were always in your face, yelling at you and looking for any possible infraction or sign of weakness, which resulted in pushups. They would get inches from your nose and shout, "What are you?"

And you'd bellow back, "Airborne, sergeant." And they would holler, "How far?"

And you'd boom back like thunder, "All the way, sergeant!"

They lived and breathed perfection, and when you lined up in the mornings for a run, or for training, your combat boots were spit-shined, your uniform was starched clean and spotless, and you'd better not have a loose thread anywhere on your fatigues. Your face was shaved clean and so was your head. Twenty-five to thirty-five pushups were counted off right then and there for each infraction, and there was no letup until the sun set.

They didn't just put us in an aircraft and say, "Jump." That would come soon enough. Instead, the training leading up to jumping from the aircraft was intense and began with the parachute landing fall (PLF) from six-foot platforms onto sawdust. We had to execute each attempt perfectly or run an extra mile or work off our mistake with pushups. Then we did the same from six-foot platforms onto hard ground. After that, we leaped from a thirty-four-foot tower on a cable angled to the ground. Then it was on to the 250-foot towers where we were each lifted in a chute and then released with the chute already open. I remember hanging there in the chute, high on the tower, and I could see all of Fort Benning and parts of Georgia and Alabama. It was mesmerizing, if not a little scary, but it sure beat the pushups. After two weeks of intense training, it was on to the real thing, jumping out of airplanes, even then the hazing never stopped.

We made five jumps from C-119s and C-123s. The first three jumps were from the C-119s, often called Flying Boxcars. They had been used in World War II and were getting old. They may have been perfectly safe, but they made me nervous. Not only did they look old—they shook, squeaked, rattled, groaned, and vibrated to the point that I was glad I had a parachute. Jumping out of them was almost better than flying in them.

Four out of the five jumps in Airborne School were from an altitude of 1,250 feet. The landing felt about like jumping off a seven-and-a-half-foot platform with no chute. Of course, wind velocity also created horizontal speed in addition to the vertical speed. We were well trained to reduce the horizontal speed and the resulting force with which we hit the ground by "slipping" into the wind—pulling the risers opposite the horizontal direction of movement. It was an important lesson that would later be reinforced to me on my second jump.

The aircraft were large, and we jumped in groups called "sticks." A stick was the maximum number of paratroopers that could exit a door of an aircraft while the aircraft was still above the drop zone. In that way, every paratrooper had a reasonable expectation of hitting the drop zone.

I will never forget my first jump. We sat on long benches against the inside fuselage of the aircraft, packed in tight. There had been so much training, in increments, that I wasn't particularly nervous. Then, once we were in the air, the crew chief opened the side door, and all of a sudden I felt the rush of wind and heard the deafening roar of the engines so close. Then it was real, and that's when it really hit home that I was about to jump out of an airplane. Up until then, it hadn't seemed certain, and yet I always knew it was.

The goal was to get everyone out the door of the aircraft as quickly as possible while the aircraft was still above the drop zone. It meant a trooper went out the door once every second. If not, the paratroopers would land spread out miles beyond the drop zone or land in trees along the periphery. In combat, a company commander or platoon leader needed to assemble his men quickly, and having troops spread out widely inhibited that.

Today it is not unusual for a ninety-year-old grandmother to make a jump with an instructor. But when you have a brigade of five thousand paratroopers in the air at the same time in a small space, and you are putting one trooper out the door of each aircraft every second from many Flying Boxcars traveling 130 miles an hour, it dials up the risk factor.

There were thirty-four men in my stick, and all eyes were on the jumpmaster, who was standing by the open door. He gave the commands and whatever he said, we did, and we did it in unison.

"Stand up." We all stood simultaneously, banging our combat boots on the cold metal floor of the aircraft in unison.

"Hook up." We snapped our static lines from our main chute to the cable that ran overhead and all the way to the door of the aircraft.

"Check static lines." I double checked my static line and made sure it was properly hooked on to the anchor cable. If not, the chute wouldn't open.

"Equipment check." I checked to ensure that everything on my two chutes was secured properly. Starting at the rear of the stick, each paratrooper checked the equipment of the jumper behind him before turning around and allowing the next jumper to inspect his gear.

Then came the jumpmaster's command, "Sound off for equipment check."

Each paratrooper sounded off one at a time, starting at the rear of the aircraft, "One okay," "Two okay," and so on.

"One minute!" All eyes were on the red light that would soon turn green.

The door was open. I was only a few feet from it. I felt the massive rush of air and the pungent smell of the C-119's exhaust.

My heart began to pound. I could hear it beating, and I felt the adrenaline surging through my veins. It was surreal, but I knew it was very real.

"Thirty seconds!" We were coming up on the drop zone.

"Stand in the door." Suddenly, like I was being controlled by some outside force, my body started moving in unison with the paratroopers in my stick as we did the airborne shuffle toward that open door, the sound of our combat boots making a loud mechanical, warlike, and aggressive sound on the cold metal floor of the aircraft. At this point I was part of something bigger than myself, and I knew the part that was bigger was going to jump.

I was number two in my stick. The trooper in front of me was standing in the door with his hands against the outside of the aircraft on both sides. It is important to leap out aggressively to avoid being hit by the rear of the aircraft or for the chute to snag on the airplane. You can probably imagine how that might turn out.

At this point I felt the shocking realization that my next two steps would take me out of this airplane.

I saw the red light turn to green. We were over the drop zone. "Go."

The guy in front of me vanished. I slapped my hands aggressively against the outside of the aircraft on both sides of the open door and leaped hard.

It was less than a second, and I was out the door.

The cadre had warned us many times about the prop blast, but nothing in the training prepared me for it, the 130-mile-per-hour shocking wall of wind or the deafening noise of the engine so close. I felt as if I had jumped into a great storm.

There was no sensation of falling, at least not for the first few seconds. This wasn't like I expected. It was more like the violence of a tornado had sucked me right out of the door.

Then I counted, "One thousand." The prop blast blew me completely horizontal. It caught me off guard. For some reason during training, I had always assumed I would be falling feet first.

"Two thousand." The force of the air ripped at my eyelids, and I kept them clamped tightly shut.

"Three thousand." I opened my eyes, and I could see the first jumper in the air ahead of me, and on each side were many others from aircraft nearby, some too close for comfort. My chin was tucked into my chest. My hands grasped both sides of my reserve chute, ready to pull the handle if my main chute failed to open.

"Four thousand." I felt the tremendous jolt that I had hoped for, and with a feeling of instant relief, I looked up. Then I realized the chute was not fully open, as the risers were twisted. I don't recall being at all unnerved. I had been trained for it, and I began kicking my legs aggressively. Kicking my legs caused me to spin in the opposite direction of the twists, so the chute finally opened fully, and I could control direction. My training had been effective.

I remember looking above me and seeing the C-119 fly away. It felt weird to see my lifeline disappear. Then something happened that I had never experienced, was not trained for, and never expected. It was surprising.

It was total silence.

The ground was too far below to hear any noise from it at all. Nothing. And there was no noise from the wind because the parachute moved with the wind at the exact same speed. I had never experienced anything that quiet. The contrast between the deafening roar of the engines and prop blast moments earlier and the total silence I was now experiencing was astonishing.

It was too heavenly for words, and so smooth. I felt apart from the world, not sure if I was even moving at all, absorbed in the moment.

And at twelve hundred feet the earth looked different than I had ever seen it, not at all like standing on the peak of a tall mountain or looking through the window of a commercial airliner. I was stunned. Spellbound. Then I remembered the sky was filled with thousands of paratroopers and other aircraft, and I had better be careful not to collide with one of them. I had to focus.

The danger was running into another trooper in the air and getting the chutes twisted together so they didn't fully open. There was also the possibility of coming down fast on top of someone's chute below, which would steal all the air from the chute above, causing it to collapse. In this case, the paratrooper above had to run across the top of the chute below and jump again, which hopefully refilled the upper chute with air. I was not eager to test the theory of this.

I looked around to ensure no other jumper was too close to me or under me, and to keep one eye on the drop zone. Sure enough, there was a guy coming in close and fast on my right. He was way too close. I slipped left, but now there was someone just below me. Quickly I went farther left and realized there was a guy coming in fast behind me, so I slipped quickly to my front. While all this was happening fast, I kept glancing constantly at the drop zone below. I wanted to make certain I hit it. Indeed I did hit the drop zone, and with a good parachute landing fall. It was surreal. I was on the ground again, and I felt a great rush. It was euphoric.

I thought the rush would end after a few minutes, but it did not begin to subside until much later, when we were in the trucks and on the way back to the airfield. We laughed and shared stories and thought maybe we were close to becoming paratroopers.

The second jump, with rifles and full combat gear, was more difficult. We were quickly gaining confidence, but it was windy that day. I didn't slip into the wind quite enough and almost buried my steel pot (helmet) in the ground upon landing.

I learned my lesson, slipping into the wind and tucking my head to my chest aggressively during the parachute landing fall, and I never let that happen again.

Then came the third, a night jump, where everything was even more risky. It's more difficult at night to avoid collisions with other paratroopers in the air.

The fourth jump was a daunting low-level jump from an altitude of only six hundred feet, which left only two seconds to pull the handle on the reserve parachute if my main chute failed to open. Otherwise, the reserve chute would not have time to fully open before I hit the ground.

Then came the final night jump in full combat gear with weapon and duffel bag, which was the most difficult and technical of all. With a hundred pounds of combat gear, the rate of fall was much faster. It was important to release the heavy duffel bag just before impact to reduce the rate of fall back to normal.

These were battalion and brigade-level jumps with hundreds of troops exiting many aircraft above the jump zone simultaneously. The entire sky filled with aircraft and parachutes. It's a sight not many have ever seen in person. But the sight never leaves you. It's that powerful.

Even with five jumps, no one was a paratrooper until the silver airborne wings were pinned on their chest. That happened immediately following the final night jump at a Prop Blast Ceremony. It was sort of bizarre.

They loaded us in trucks in the darkness of the drop zone and took us to the ceremony location in the boonies. There, in the middle of the night, seated behind a small folding table in a remote field, was the Fort Benning commanding general, the commanding general of the Jump School, and the sergeant major of the Jump School. The table faced a six-foot-high wooden jump platform with steps leading up to it.

To pass the Prop Blast Ceremony test, each paratrooper had to drink a 155-mm howitzer shell full of bourbon, vodka, and scotch. A howitzer shell is very large, and an empty one holds a lot of booze. After drinking that powerful and large concoction, one by one, we climbed the steps of the six-foot platform and had to then jump and execute a flawless and simulated parachute landing fall on the ground in front of the committee. The generals and the sergeant major gave us a thumbs up or thumbs down depending on how well we executed the parachute landing fall.

If you got a thumbs up, it meant you passed, and they pinned the silver airborne wings on you right then and there in front of everyone, and you were, at last, a paratrooper.

But the troopers that got a thumbs down had to completely drink a second howitzer shell full of the same alcoholic mix. Then on wobbly legs they climbed the steps once more and executed another parachute landing fall, and so on until they finally got a thumbs up. Some were so drunk they never remembered the silver airborne wings being pinned to their chest.

I was lucky and passed on the first attempt. But many did not, and more than two-thirds of the men were completely stone-cold passed out. I helped load their limp bodies onto the trucks in the darkness to take them back to the barracks, where I helped unload them into their bunks.

The whole Prop Blast Ceremony was something the Army later discontinued because they had some deaths from the alcohol intake and took a negative PR hit. Years later, the graduation ceremony became much different. Guests and family were invited to attend a formal and "politically polite" graduation ceremony, sitting in bleachers. Those same families would have been shocked to have seen what our Prop Blast Ceremony was like or to have watched their sons hauled off in a drunken stupor.

Nevertheless, and regardless of how the ceremony was conducted, the result was the same. We were paratroopers and proud of it.

UNBROKEN, 1965

NEXT CAME THE biggest challenge I would face in my military training: Ranger School. If paratroopers were a step above infantry, the Rangers were a giant step above paratroopers. The Rangers were an elite force within the army and looked up to by many. At the age of twenty-two, with my new silver airborne wings proudly displayed on my chest, I reported into Ranger School, also located at Fort Benning.

The unstated goal of the staff at the Ranger School was to ensure that less than half of the class made it to graduation. Usually only 40 to 45 percent made it through. So out of a class of 175, around 70 to 80 became Rangers. Only soldiers within the army who were in the best physical shape even bothered to apply to Ranger School, or were accepted, otherwise the dropout rate would have been much, much greater.

Ranger training was designed to stress soldiers to a point just short of death. Being "just short of death" was hard to define. Some deaths occurred through the years in Ranger training as it was very difficult to estimate where the point just short of death really was.

It was said within the military in those days that the stress of Ranger training was equal to one full campaign of actual combat. And in fact, in 1965, most of the Ranger cadre were veterans from the Korean War. They demanded relentless tenacity in the face of adversity, perfect decision-making under extreme stress, and complete mastery of ambush and recon patrol skills behind enemy lines or in a guerilla warfare environment.

I had a short break between Airborne School and Ranger School. Naturally, I wondered what Ranger training would be like and how to prepare for it. I had just finished a long run on Eubanks Field late one evening to get in top shape. I completed my run at the parking lot at the same time as another soldier. Even though it was almost dark, I could see he was wearing a Ranger cap. I had never had an opportunity to talk to a real Ranger one on one, so I walked over and introduced myself.

He looked to be about four or five years older than me, with a lean and hardened look, muscular and chiseled like a soldier who had spent his whole army career in the field. We chatted politely for a minute, and then I asked him about Ranger training and how I should prepare for it. He smiled knowingly as if he had heard that question before.

He had a pint-sized bottle of Early Times in the trunk of his car and asked, "Would you mind if I took a sip while we talk?" He offered me some, and I declined. Then, he leaned against his car in the encroaching darkness and said, "There are only a few things you can do to prepare for Ranger training. The most important is what you're doing now, to be in top shape. But strength and athletic endurance will only take you so far. There's a whole lot more in Ranger training that you can't possibly prepare for."

He looked at me for a long time as if he were sizing me up, and finally he continued, "The worst of it is lack of sleep. You'll go three days and nights on long range patrols with no sleep at all, walking nonstop the whole time." He added, "I've seen some very tough men turn into zombies—seen 'em fall asleep while walking in the darkness and remain asleep even as they crash into trees, stumble over logs or fall down the side of some steep mountain. You can't prepare yourself for lack of sleep. Nobody can. It's not even possible."

I was still absorbing this as he added, "It's three months of starvation. I mean severe starvation. I can tell you for certain, there is absolutely no way you can prepare for starvation."

"When you're walking day and night for three days with no food or sleep—when you're dog-tired and so weak from hunger that you can't even think, and you're ready to drop, that's when the Ranger cadre will push you the hardest to see whether you can lead a patrol and make the right decisions. It's a combination of fatigue and stress that's beyond anything you could ever imagine or dream up on your own, much less prepare for."

Then he looked at me closely, as if he were once again sizing me up, and he said, "All the weightlifting, pushups and chin-ups, sports and physical exercise you've done in your life and the runs you're doing now to be in top shape—they build nothing but muscle strength. Don't get me wrong, you'll need it—all of it, and then some. But there are challenges that come at you in Ranger training—adversity like you never knew existed, and after three days with no food or sleep, it will rob you of all that muscle strength. Not just some of it, all of it, and the only thing that's left is what's inside of you, the stuff you're made of. It's called grit, and no one knows how much of that grit they've got, and whether it's enough, until it happens."

Then he added, "I wish that I could tell you something more to help. But I can't because Ranger training is about finding out what's inside of you and how much adversity you can take. And trust me, making it through Ranger training has a lot to do with how much you want that Ranger patch on your shoulder."

He wished me luck, and I could tell he meant it. Then as he was about to drive away, he rolled his window down and said, "I nearly forgot one of the most important things. It's about the Ranger cadre. The thing they care the most about. One burning question. Which Ranger candidates would they want to go into combat with—to be with on an ambush patrol behind enemy lines?" Then he drove off into the darkness.

It was a long way back to my quarters, and I couldn't help but wonder what I had gotten myself into—how much grit I had inside of me and whether I would measure up.

There were 175 who reported in with me that cool crisp morning at Fort Benning, and we were assigned to several of the old wooden barracks at Sand Hill. I thought I was as prepared as any man could be, but even with all the forewarning, my first moment in Ranger training was a total shock. The hazing and the pushups were beyond intense. It was insane and nonstop, twenty-four hours a day. The Ranger cadre were intent on breaking us to the point where we would quit.

The training was just as intense as the hazing: hand-to-hand combat, weaponry and patrolling at night, heavy physical conditioning many times daily, and nonstop rope climbs, PT courses, obstacle courses, and multiple five-mile runs daily, some at midnight, some by day, and some at 3 a.m. in combat gear. As a matter of fact, Ranger candidates ran everywhere, and the running was a lot faster than the airborne shuffle in Airborne School. Woe to the poor Ranger candidate who was ever caught walking anywhere, even for a second, or not running fast enough. The only time we walked was on patrols. And patrolling was the real strength and role of a Ranger unit.

The most important training was on how to lead patrols behind enemy lines. The emphasis was on recon and ambush patrols. There was advanced map reading (all pre-GPS) and compass work (beyond what was taught in Infantry Officer Basic Training). Rangers were expected to know where their patrol was within ten meters at all times and be able to call close-in friendly mortar and artillery fire without hitting their own men. On each patrol, you were graded, but the real test the Ranger cadre used was simple—who were the soldiers you'd want with you if you were going into combat?

Forty Ranger candidates dropped out or were culled in this first phase of Ranger training at Fort Benning. Out of my original class of 175, we entered the mountain phase with 135 still intact.

The second month, or mountain phase, was heavy on explosives, physical training, more intense hand-to-hand combat, and significantly longer patrols through the rugged mountains of North Georgia in small units. At first we were given one C-ration meal per man to pack with us, but one meal for three days of nonstop day and night hiking up and down steep mountains is not sufficient, and frequently we caught and ate snakes as well as watercress, ramps, and other plants from the forest.

We learned how to recognize enemy ambush sites, how to fight our way out of an ambush, and how to climb and rappel cliffs at night. When we were not on patrol, the multiple five-mile runs daily got progressively faster and more frequent.

In addition, there were combat courses where we crawled quickly under barbed wire for hundreds of yards with machine guns firing live rounds just above our heads. I was the first man through one of the courses that hadn't been used for a while and had grown up on the bottom with heavy poison ivy. I had taken my shirt off so I could slither low under the barbed wire on my belly like a snake, staying close to the ground without my shirt hanging up on anything. Too high, and I would be close to the bullets flying over overhead. Too slow, and I would be graded down on points. I had to be fast and low.

I hadn't counted on the poison ivy; and I am extremely allergic to it.

After slithering on my belly with my rifle for two hundred yards through continuous poison ivy, I immediately broke out in an uncontrolled and growing rash. My arms were swollen to over twice their normal size. I lost control of my wrists. My body looked nightmarish, with huge red and pink boils, rashes and massive swelling.

The Ranger staff gathered around and proclaimed this would knock me out of Ranger School. But I hadn't come this far not to be a Ranger. Surely there was something I could do. I convinced them to send me to the infirmary, where there was a doctor who examined me. He said there was a chance he could help, but that it was a long shot.

That turned out to be literal: He pulled out a huge syringe that looked to me to be about twelve inches long and filled with what he said was cortisone. It was larger than any syringe device I had ever seen or heard of by orders of magnitude. Not even in fictional or horror movies had I seen one that big. Then he smiled and said, "Turn over on your stomach and I'm going to shoot the entire contents into your left buttock."

There was a tremendous sting like a hornet or wasp, and it lasted a while.

Then the most miraculous thing began to happen. Over the next two hours, the poison ivy swelling subsided. By the end of the second hour, it had almost completely disappeared, and I was back in my Ranger unit, ready to go—to the total amazement of the other Rangers and the cadre.

Someone was watching out for me on that day. I thank God and wish I could also thank that doctor. I have no idea who he was.

During the remainder of mountain phase, we operated mostly on two- and three-day patrols. There was one particularly grueling three-day patrol through some particularly steep and rugged mountains with no food or sleep at all. We kept moving the whole time. A lot of the dropouts and culling in the mountain phase came from that long patrol. By now, seventy of the original Ranger candidates were gone.

The final month was the swamp phase in Florida. We entered the swamp training with 105 Ranger candidates still intact from the original class of 175.

We were awakened the first morning of swamp training at 3 a.m. by an NCO (non-commissioned officer) Ranger named Sergeant Parker. He was built like Jim Brown, one of the greatest football players in history, back when he played for the Cleveland Browns. Sergeant Parker had that same rugged and physical build.

His first words on lining us up in the darkness were "Rangers, this is going to be the worst run of your life."

We were in formation on a dark and lonely dirt road leading through the swamp, and he said, "Look at the man on your left. Now look at the man on your right."

"Half the men you are looking at will not be here at the end of this month. This run will knock a lot of you out of Ranger School. In fact, about half of you that don't finish Ranger training will be culled on this run. This will be the fastest five-mile run of your life."

"Look at your buddies at the end of this run who finish near the front. They will be the Rangers who make it through swamp phase. Let's go."

I had no idea how fast this Ranger was going to run. I knew this would be a defining moment as to whether I became a Ranger. It was faster than anything I had seen in Ranger training, and the farther we went, it just got faster.

I worked my way in close behind him, determined never to let him get more than twenty feet ahead. After four and a half miles he went into a moderate sprint. It took everything I had to keep up with him. But I sprinted with him and ended up almost right beside him. I was the very first Ranger candidate he spoke to at the end of the run. He simply turned to me and said, "You'll make it."

Soon he was taking the names of the poor stragglers.

And of course, he was right. I looked at the Ranger buddies of mine who finished in the top half of that run. For the most part, they were the ones who ended up making it through swamp phase and becoming Rangers.

Swamp training was primarily three-day foot patrols through waist-deep swamps as well as waterborne operations, small-boat movement, and stream crossings.

The emphasis was on the skills needed to survive and accomplish effective recon and ambush patrols behind enemy lines in a rainforest or swamp environment.

There was training on venomous snakes, alligators, and swamp varmints of all kinds, including how to cook them, which seemed odd, because we had already eaten snakes on the longer patrols, even since the beginning of mountain phase.

The worst day of my life was the day the cold front swept in from the north. It was one of those freak cold fronts with a strong north wind and bone-chilling, nonstop rain. By the third day of the patrol, with no food or sleep, we were soaked to the bone and beyond exhausted. There, in the darkness of the swamp, we were given the coordinates to a rubber raft hidden in a friendly cache. We had to locate the cache site miles away through dark, waist-deep water. Once the raft was secured, we had to paddle it from the swamp and across a windswept bay, carry it across a barrier island at night, launch it in the heavy surf of the Gulf of Mexico in this bitterly cold and driving rain, and then paddle five miles out into a seriously rough sea to locate a landing craft with no lights for linkup.

By the third or fourth mile out into the darkness and cold of the Gulf of Mexico, with the north wind howling, the rain in a frenzy, and the waves dangerously high, most of us were in the early or intermediate stages of hypothermia. Each wave broke over the small raft and drenched us, but we stayed the course. And at the fifth mile, we found the landing craft in the dark with no lights. We were dead on course with no landmarks to go by.

There was no cover on the landing craft, and we lay there in our wet fatigues on the exposed, cold metal deck in the wind and rain shivering so badly that we lost all control in our hands and arms. Some of us, including myself, were indeed approaching that "point just short of death."

We lay there in the cold north wind and rain for nearly an hour while the landing craft searched for and linked up with a larger Navy patrol boat, also with no lights. The patrol boat had a heated cabin, and the Naval officers and enlisted men cranked up the thermostat and fed us hot cereal, coffee, and doughnuts. Ever so slowly, our substantial shivering began to lessen, and after a half hour or so, we regained control of our bodies. We had been three days through a storm in the swamp with no sleep, and it was our first food in all that time.

Graduation was back at Fort Benning.

Only seventy-nine of the original 175 had made it through unbroken to became Rangers. Most had lost thirty pounds or more during the three phases of Ranger training. Maybe I was in better shape to start with, but I went from

180 at the end of Airborne School to 160 at the end of Ranger training for a drop of only twenty pounds.

In an elaborate ceremony at Victory Pond back at Fort Benning, the black-and-gold Ranger tab was pinned to my left shoulder by the Ranger battalion commander. The Ranger tab is permanently worn above the unit patch on the left shoulder and moves to the right shoulder once the soldier serves in actual combat.

My family was there and proud as always. For me it was the realization of a life dream—a dream that had started when I was just a little kid.

The top Rangers in our class got choice assignments. I chose to be a rifle platoon leader in the 82nd Airborne Division at Fort Bragg, North Carolina, in Fayetteville.

I reported to Fort Bragg, was assigned to lead a rifle platoon, and was there for only two days when they asked for volunteers for a special airborne Ranger unit going to Vietnam. I volunteered, as did many others.

Four hours later, I had my orders to report the next morning to Fort Benning, Georgia, to the First Battalion, Eighth Cav, First Brigade Airborne, part of the First Cav Division on its way to war. I was assigned to be the First Platoon leader in Company C (Charlie Company). My training was done at last. I had made it through unbroken. And now the real test was about to begin.

N
W
E
S

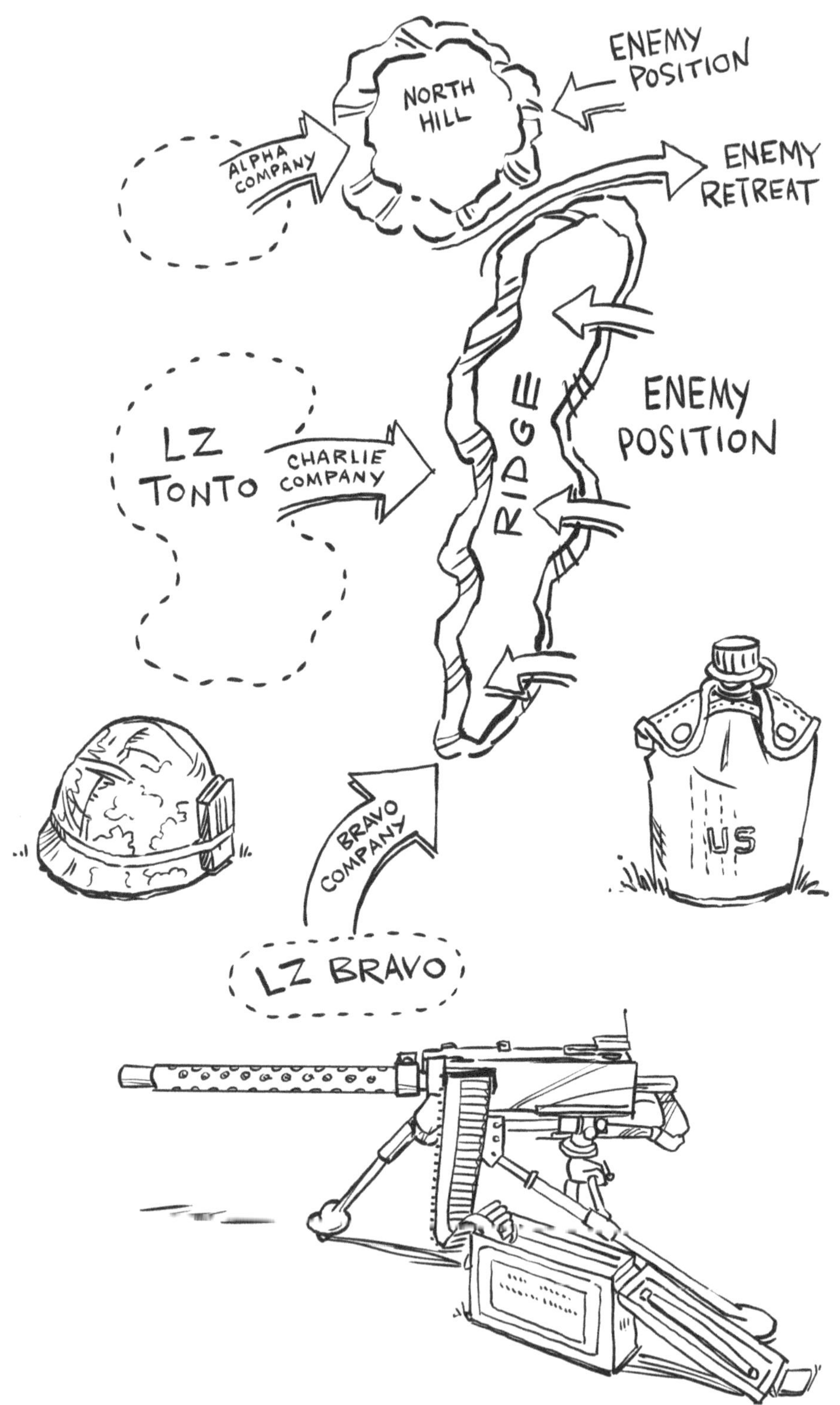
ENEMY POSITION
NORTH HILL
ALPHA COMPANY
ENEMY RETREAT
ENEMY POSITION
RIDGE
LZ TONTO
CHARLIE COMPANY
US
BRAVO COMPANY
LZ BRAVO

THE BATTLE OF HAPPY VALLEY, DECEMBER 18, 1965

ON THE EIGHTEENTH DAY of December in 1965, there was a battle. A battalion of Americans squared off against a battalion of Viet Cong (VC) and NVA.

No movies were made of it. There was no coverage in the American press that I was ever aware of. In fact, the American public had never heard of Happy Valley. But if you asked anyone in the First Cav Division, they could tell you stories about the place. Not because Happy Valley was a happy place: quite the opposite. And certainly not because they had been there. No one (at least no one that we had ever heard of) had ventured into the heart of Happy Valley.[4] Only a few had even dented the periphery. It was simply thought of as a mysterious and dangerous place, completely in the hands of the VC and NVA.

The area we called Happy Valley was in Binh Dinh Province, north of highway 19, the only east–west highway through the central mountains of Vietnam. It was our supply route between our First Cav headquarters at An Khe and our supply base at Qui Nhon on the South China Sea.

The VC and NVA had always controlled Happy Valley, even as far back as the French Indochina War. From this sanctuary, they attacked our supply chain to the outside world.

4. For context and clarification, American soldiers sometimes called a place controlled completely by the enemy a Happy Valley. The Marines later named another one to the north near Da Nang.

On the evening of December 17 at 2100 hours, Captain William Smith, our company commander, called the platoon leaders of Charlie Company to the command post and gave us the warning order. Captain Smith had been our company commander since the day I reported in at Benning, and he was a good one—a great officer and well respected. He told us that we would air assault into the heart of Happy Valley with the objective of taking a hill and ridgeline where a concentration of enemy troops had been spotted from the air the previous day.

My platoon (First Platoon) would lead the way for Charlie Company at the head of the first wave of choppers, assaulting into an area we called LZ (landing zone) Tonto, on the west side of the ridgeline. The helicopters would then return to An Khe and pick up Bravo Company in a second wave, followed by third and fourth waves of Alpha and Delta companies. There would be artillery prep in advance on the objective and a pre-strike by the Air Force.

Intelligence reported an enemy presence of unknown strength on the ridge with several battalions of Viet Cong and NVA somewhere within striking distance. There was a possibility we would be badly outnumbered.

Charlie Company's objective was to assault and take the long ridge rising between fifty to seventy feet above the surrounding flat rice paddies, where most of the VC and NVA had been seen.

I studied the aerial photo for hours that night with the other platoon leaders. The ridge was covered in vegetation and high trees and surrounded by open fields and rice paddies. The dikes between the rice paddies were vegetated, making them appear almost like hedgerows in the photo.

First Platoon would be on the lead choppers, and we would secure the southeast perimeter of LZ Tonto, then link up with the Second and Third Platoons and be prepared to attack the ridgeline. The assault on the ridge would commence as soon as Alpha Company arrived on the third wave of choppers and had secured the northernmost hill overlooking the LZ.

The weather was closing in fast. It began raining heavily, and the storm raged on all through the night. I remember lying on my cot listening to the rain hammering my tent. Just before midnight, I had assembled First Platoon and given them the operations order, and now there was nothing to do but try to sleep. My mind drifted to a place far away. I don't know why, but for a moment I tuned the little transistor device to Armed Forces Radio and learned that the next day, December 18, the University of Tennessee, my alma mater back in Knoxville, would be playing in a bowl game. I thought it ironic. I knew exactly what my friends back home would be doing tomorrow, but they would have no clue where I was or how I would spend my day. Neither would they have recognized the name Happy Valley. It would have meant nothing to them, just as many Americans at that time couldn't have said where Vietnam was. It was the last thing I thought of as I went to sleep.

I awoke hours before daylight in a pouring rain. By 0700 the storm had tapered to a light rain, and my platoon was drinking hot coffee and working on their equipment. We had lived this scene so many times—air assaults into suspected jungle strongholds—sometimes with resistance but just as often in places the enemy had just vacated. However, today seemed different. This was Happy Valley, and there was a magnitude to this mission. It was palpable.

H-Hour was 0900 as my platoon climbed aboard the Hueys from the 229th Aviation Battalion. We lifted off in a low ceiling and light rain. I was with first squad in the lead chopper on the first wave, which included the remainder of First Platoon and the rest of Charlie Company.

It seemed like a short flight, but a lot was going through my mind. For one thing, I was the senior rifle platoon leader in the company. Our executive officer had been wounded in an earlier mission, which meant that if anything happened to Captain Smith, I would take control of Charlie Company. That would be a big responsibility for a platoon leader, and it weighed heavily on my mind.

We flew the Hueys in Vietnam with doors removed on both sides. I sat with my back to the engine facing out. The blast of air and roar of the engine were exhilarating and incredibly loud.

We flew a tight formation—very tight. I had lived this so many times and was always amazed that the chopper pilots were able to fly so close together. It always appeared as if the rotor blades of the chopper closest to us were almost touching ours. Probably an optical illusion, but it took some getting used to.

We were near the LZ when somewhere up the chain of command they decided the weather precluded a pre-strike by the Air Force. But of course, the artillery prep would help us secure the LZ.

At we approached the ridgeline, the pilots suddenly began flying erratically, swerving violently from side to side. I thought that was odd, in such a tight formation. Then I understood. We were taking fire from below.

I didn't know it at the time, but the artillery had also been diverted at the last minute to support Third Brigade in a firefight to the south. We were headed into a hot LZ in the heart of Happy Valley with no prep from either the Air Force or artillery.

With no prep, the NVA had absolutely no idea we were coming until we were right on top of them. Normally, the element of surprise in combat is a strong advantage—unless you are a rifle company about to land right in the middle of a battalion of NVA and Viet Cong, which was exactly at that moment what was happening.

At about fifteen seconds out, our chopper was hit by fire from the ridge below, but no one was wounded and the chopper was okay. From the corner of my eye, I saw another chopper swerve dangerously close. I thought it would crash into us, but it swerved just barely underneath us at the last second. Other choppers were lurching and veering erratically, and it was chaos in the air. Then we were over the LZ.

The bird I was on was the lead chopper. The pilot never got closer than eight feet from the ground, that way he could keep the pitch on the blades and get out of there faster. I was hanging outside the chopper with my feet on the skids to create less vertical distance on the jump, but the ground where I would have landed was being torn to shreds by machine gun fire. I could see the ground being ripped up every few inches. Everyone else in first squad saw it too, because I hadn't yet seen anyone on my Huey jump.

For a split second, the rounds stopped hitting the ground, and I leaped from the chopper, rolled on the ground and came up running toward the southeast side of the LZ. As far as I know, I was the first American on the ground in Happy Valley. I could hear that cracking sound that rounds make as they come by your ears, too close.

I looked around, thinking the rest of first squad was right behind me. But they were pinned down right where they had jumped from the chopper, and the entire LZ was being ripped with machine gun fire and 60-mm mortar rounds.

One of the choppers was hit and disabled on the LZ by enemy gunfire. I saw it off to my left on the ground, but it looked to me like all the troops were getting out of it okay.

I could already see three or four of our guys wounded and others pinned down in the open with little cover. During the confusion of landing under fire, some of Vavrek's Third Platoon had gotten mixed in with First Platoon. There were wounded from both platoons.

As I moved to help organize, and try to get a medic to the wounded, enemy snipers and machine gunners opened up, causing more casualties. So far, I didn't see or know of any dead from our platoons, but I could see less than half the LZ from where I was.

Captain Smith called for supporting fire, and two ARA gunships from the 229th swooped in from my right (the south) and made a run on the ridge with rockets and machine guns. That suppressed the fire from the ridge for a moment. It gave us valuable seconds to begin organizing squads and help our wounded men to the edge of the LZ where the cover was better. It was mostly just two-foot grass where I was, but there were also a few little humps and mounds here and there that provided partial cover.

The noise of gunfire on the LZ was deafening as the gunships circled around for a second strafing run on the ridge. They were coming in low and fast from my right. One of them was hit, and I watched it go down, disappearing near the southwest end of the ridge.

The guy to my left was hit and lying in the open. One of the medics was near me, and we half crawled, half ran over to him, crouched and low. He was from one of the other platoons, and I had never seen him before. He had a bullet wound in his upper body but was conscious. We were lying there as low as possible with rounds hitting around us, and I remember he said to me that he was okay and could still shoot. He asked for a cigarette. The medic was giving him morphine and also gave him a cigarette. When he inhaled, the smoke came out of the wound in his chest. The sight of that was awful, but I kept my face neutral so he wouldn't know how bad I thought his wound was. He was alert and brave and wanted to continue fighting, though I knew he could not. The medic taped him up, and with enemy rounds hitting all around us, the two of us helped move him to the edge of the LZ where there was slightly more cover.

I was on the radio requesting a medevac chopper, as were the other platoon leaders, but the LZ had already been deemed too hot for medevac to land. It was a tough situation. Charlie company was outnumbered and exposed on the open LZ, fighting a battalion of VC who were dug in on the northern hill and the entire ridge.

As the firefight raged on, the choppers had returned to the base and picked up Bravo Company under the command of Captain Roy Martin. Captain Smith diverted Bravo Company's landing from LZ Tonto (where we were) to a new LZ to the south (LZ Bravo). That way, Bravo Company could assault the southern end of the ridge in a coordinated attack with our company from the west and perhaps even rescue the downed ARA gunship pilots if they were still alive.

I watched as the remaining ARA gunship marked the new LZ Bravo with yellow smoke so the choppers carrying Bravo Company would land at the right place.

To my right, I saw the choppers carrying Bravo Company under the command of Roy Martin. They came in low from the south in the spitting rain. Captain Martin was a veteran of the Korean War and a former commander of the vaunted US Army Golden Knights Parachute Team. He was a smart, tough soldier and respected by everyone.

From where I was, I couldn't see Bravo Company land as the LZ was slightly blocked from view, but I could hear the intense barrage from the VC on the south end of the ridge. We were placing the highest possible base of fire on the ridgeline to support Bravo Company's landing, but they were taking heavy casualties.

Suddenly, in this melee and bloodshed, another chopper came flying in. It was attracting heavy machine gun fire from a bunch of locations on the ridgeline. Nevertheless, the bird landed to my left and about a hundred yards away.

At first I thought it was a medevac chopper, and many of us were helping wounded men toward it. The crew chief was also helping load the wounded, but he was killed by NVA fire. As it turned out, it was not a medevac at all. It was our battalion commander's chopper. Circling overhead and running low on fuel, he had come in to command the fight from the ground and allow for some of the wounded to be evacuated.

As many wounded men as possible were loaded onto the chopper—so many, in fact, that the bird barely got off the ground. I wouldn't know until later, but it made it back to our base camp at An Khe with the gas gauge on empty and twenty-two holes in the aircraft, including one in the compression chamber of the engine.

The firefight raged on as we waited for Alpha Company so we would be full strength before we initiated the attack. In the meantime all platoons of Charlie Company continued to take casualties.

At last, the first medevac chopper arrived, supported by the remaining gunship. The medevac bird attempted to land to my left. I watched it come in under heavy fire, and as it touched down, the pilot lifted off and flew away. The copilot had been killed.

One of the platoons found a depression northwest of Tonto that would provide some protection for the medevac choppers to land, and all three platoons began moving their wounded across open ground to this more protected location and very soon the medevacs arrived.

Meanwhile, enemy automatic weapons and mortar fire were covering pretty much the entirety of LZ Tonto. The VC were well dug-in on their higher ground, while here on the open and exposed LZ Tonto, we were more visible. This wasn't the kind of behind-enemy-lines fighting that I had been trained for as a Ranger. This was conventional warfare much like World War II, and it was going to take a direct assault on the hill and ridge before the battle could be decided.

The firestorm continued unabated as the choppers carrying Alpha Company came roaring into Tonto under command of Captain Ted Danielson. He was a West Point graduate from Columbia, South Carolina, and a tough but well-liked officer. His company's mission was to attack and take the open hill to the northeast. Once Alpha Company secured the hill, they would then lay down a base of fire to support our attack on the ridge—Charlie Company attacking from the west and Bravo from the south.

We poured a base of supporting fire on the exposed northern hill while Alpha Company initiated its attack. Due to the openness of the terrain on the hill, I had a clear view of Captain Danielson's platoons maneuvering up the open slope, one platoon laying down a base of fire while other platoons advanced. Finally, we had artillery support, and Captain Danielson was walking a barrage in front of his platoons as they moved forward up the hill, white phosphorus rounds exploding repeatedly.

Even with the artillery support, Alpha Company was taking heavy machine gun fire from tunnel openings. It was on that hill that a platoon leader and friend I had known from Fort Benning, Lieutenant Stuart Tweedy, was killed by one of the enemy machine guns.

Stuart was well liked and respected, having previously distinguished himself in a nighttime air assault to reinforce a beleaguered patrol base in the Ia Drang Valley. He was a great platoon leader, and he led his men up that hill under direct enemy fire.

At 1220 hours, as Company A secured the top of the northern hill, the rest of the heavy weapons company (Delta Company) arrived where I was on LZ Tonto. With Alpha Company on the northern hill and the arrival of artillery and mortar support, there was now a tremendous base of fire directed at the ridge.

At 1300, we began our coordinated attack, Charlie Company from the west and Bravo Company from the south. As Charlie Company advanced, first platoon delivered a base of fire to support the advance of second and third platoons. The VietCong and NVA were now being hit by mortar, artillery, and supporting fire from Alpha Company as well as from First Platoon of Charlie Company. Nevertheless, the barrage from the ridgeline was intense, so much so that third platoon leader, Lt. Vavrek, took a round through the front of his shirt, one through the back of his shirt, and one through a poncho tied to the back of his belt. Miraculously, he was never scratched.

I couldn't see Bravo Company to the south, but I heard the heavy gunfire as they assaulted the southern part of the ridge and drove north, rescuing the pilot and copilot who were both still alive.

Meanwhile, we advanced from the west through rice paddies and open areas in front of the ridgeline where some of our casualties occurred. It was in the middle of one of those open areas when I heard a "swoosh" and was immediately lifted off my feet and slammed back to the ground as I heard the explosion just behind me. My RTO (radio telephone operator) was beside me on my right, somehow unscathed. He helped pull me up. I was dazed but okay.

Meanwhile, down at the south end of the ridge, the NVA mounted a counterattack against Bravo Company. They fought if off, killing thirty VC and capturing one. Then Bravo continued their attack from the south up the end of the ridge.

I think that all of us fully expected the heaviest fighting to be up higher on the ridge. However, the farther we advanced, the less enemy fire we encountered due to the arrival of artillery and mortar support as well as the intense supporting fire from the north hill by Alpha Company. The NVA had tunnels built from their firing positions on the ridge which led to escape routes on the backside of the ridge.

As we crested the ridge, I slid into a large artillery crater along with my RTO and two guys from my platoon. Soon, we were joined by a couple of guys from the advance element of Bravo Company. I remember saying to one of the soldiers from Bravo, "Hey, you guys had it pretty rough down there at the south end of the ridge."

The guy just smiled and said, "Lieutenant, it looks like you had it pretty rough too."

I asked him, "How do you mean that?"

He said, "Sir, have you not looked at your helmet?"

I removed my steel helmet, and the entire right side was gashed from front to back. I remembered the explosion behind me and being lifted in the air and slammed back to the ground.

The helmet was scary to look at. A half inch to the left, and I would not be writing this story. Maybe it was dumb luck. Or maybe it was not part of God's plan for me to die on that day. Or maybe both.

We searched the battlefield and began to realize that the VC fortifications were heavier on the west side of the ridge overlooking LZ Tonto where First Platoon and the rest of Charlie Company had first landed. The enemy had clearly expected any attack to come from there.

Fortunately, Bravo Company landed to the south, causing the NVA to fight in two directions. Otherwise, we would have sustained much higher casualties. Our commanders had made good split-second battlefield decisions for that to happen.

Later, back at our base camp at An Khe, there was time to reflect, and it struck me how fortunate I had been. It was one of the first times in history, along with the battle at LZ X-Ray in the Ia Drang Valley, where an air cavalry company landed physically right on top of an enemy battalion. Now there was time to ponder the risks, and I thought about the ones who were wounded and the ones who died. It was a strange war, and I wondered what else it had in store for us.

I dug around in my duffel bag and found the little transistor radio, and once again, I tuned it to Armed Forces Radio, searching for that part of my life that I had left behind. I discovered that on the same day as the battle in Happy Valley, a quarterback named Dewy Warren had led Tennessee against fourth-ranked Tulsa in a Bluebonnet Bowl victory.

Somehow that news connected me to Tennessee, and to my friends and family, and for a moment the world seemed smaller—and the part of that world that I had left behind seemed closer, even though it was so far away.

The football game had been carried on NBC and broadcast to millions, while on our side of the world, there had been a battle waged with no coverage by the American press. Sixteen American soldiers had lost their lives and another sixty-five had been wounded. One hundred five NVA had died. All to take a hill and a ridge in a place called "Happy Valley."

HIGHLANDS
HAPPY VALLEY
AN KHÊ
HIGHWAY 19
TO PLEIKU & CAMBODIA
CENTRAL
TO QUI NHON
AN MY RIVER VALLEY

A WAR OF WILL, DECEMBER 10–11, 1965

TWENTY MILES SOUTH of our base camp at An Khe, in the Central Highlands of Vietnam, there was an area of dense jungle mountains and draws that fed into the An My River. Battalion intelligence officers reported that new North Vietnamese units had moved into the area.

Our mission was to conduct a search and destroy mission with the initial objective being a ridgeline with thick trees and jungle canopy. There were no clearings large enough for choppers to land. We would use rope ladders lowered from the aft section of the larger Chinook (CH-47) helicopter, permitting an entire platoon of forty-three men to clamber down a hundred feet through small openings in the tall jungle forest.

We pretested the ladders from Chinooks in base camp and discovered that the rope ladders twisted when no one was on the ground holding the bottom. It made climbing dangerous.

It was also difficult to hold onto the bottom of a hundred-foot rope ladder from the ground and prevent the twisting. The Chinook pilots were experienced, but the movement of the chopper from side to side and front to back made it hard for the person on the ground to hang on.

Naturally, all the troops needed to be climbing the same side of the ladder. The pilot had to keep the chopper perfectly steady or drifting ever so slightly forward. If the chopper drifted to the rear, the troopers would be climbing on the inverse side of the ladder with their feet dangling in the air instead of on the rope rungs of the ladder.

There was another problem as well: when our men climbed down into the jungle, there would be no one already on the ground to hold the bottom of the rope ladders. The solution was to send in a group of Rangers to rappel in ahead of time and hold the bottom of the ladders as the first wave of troops arrived on the CH-47s.

On the morning of December 10, 1965, I rappelled into the ridgetop with a small group of Rangers. The arrival of the first wave of CH-47s with troops followed swiftly and attracted only occasional sniper fire, but it was from a long way off and not accurate.

The second wave came in without a hitch, and Charlie Company set out a defensive perimeter on the high and heavily forested ridge.

Meanwhile, Alpha and Bravo companies had air assaulted into a clearing in the river valley many miles away. Meeting no resistance, they drove toward us through thick jungle. As they got closer, we started taking occasional sniper fire and incoming rounds from a 50-caliber located somewhere on one of the surrounding ridges. We were never able to locate the source of it.

Suddenly we received a new order to prepare for immediate extraction and a new mission. An hour later, the first wave of two CH-47s arrived overhead and lowered the rope ladders down though the high trees and canopy. I decided to wait and go out with the second and final wave.

When the last chopper arrived, I sent my platoon sergeant up the ladder first, followed by the rest of my platoon while I remained on the ground to hold the ladder for my men. I knew full well that this meant I would climb the ladder alone with no one on the ground to hold the bottom of it.

Almost all the men had climbed up the ladder and into the chopper, and I was still on the ground trying to hold the rope ladder for the last two men still climbing near the very top. The Chinook made a big target for enemy snipers, and the pilot must have been getting nervous, because the next thing I knew, as the last two men made it into the chopper, it was lifting off.

I started climbing as fast as I could, with arms already tired from holding the ladder for so long. Suddenly, the Chinook was out over the valley, gaining altitude and speed. I was only a third of the way up the hundred-foot ladder, and it was twisting and waving wildly, flapping violently from side to side. Climbing became almost impossible. I glanced down and saw that we were now about six hundred feet in the air and moving at a pretty good clip. The pilot must have wanted out of there fast. I guess he figured he had the lives of forty-two men already on the Chinook to think about and not the Ranger who had been holding the rope ladder for them.

I realized that with the twisting of the rope and the increasing rate of speed, if I made one misstep, I was gone. The most important thing was to be sure my hands didn't slip each time I moved up a rung on the rope ladder. I was halfway up, and the Chinook was approximately eight hundred feet off the ground and moving fast. I was trailing behind and waving wildly from side to side in the wind like a leash on a runaway dog.

Between the wind speed, the rope twisting, and the severe side-to-side whiplash of the ladder, it was a grim situation. And it was for real. I decided if I died, it would be from sniper fire or some other reason—but not because my hand slipped. I grabbed each rope rung of the ladder in a death grip as I slowly moved up from one to the next.

I could see my platoon sergeant watching anxiously from the rear open deck on the Chinook high above where the ladder was attached. I thought about Airborne School and Ranger School and all the physical training I had endured, particularly the pullups and rope work. It gave me confidence that I had overcome it. I knew if I could do all that in Ranger training, then surely I could find the inner strength I needed now. My life depended on it.

At twenty feet from the chopper, the muscles in my arms became so tight they felt immobile. I was a thousand feet in the air due to the depth of the valley below, and we were moving fast. The airspeed and the wild flapping along with the twisting of the rope made the climbing more difficult. At fifteen feet from the chopper, my arms felt like they were going to burst.

Everything became a blur, the ladder flapping erratically and wildly. It was odd, but for a moment, there was a battle going on. Not with an enemy; it was in my head—between my ears. A voice in me was saying, "Your arms are tired. You may not make it. No one has done this before. Look how high in the sky you are."

And another voice in me said, "This is the fight of your life. You can do it. This is why you've stayed in shape. You're a soldier. A Ranger. Just focus. One rung at a time. You can do it. You have to."

I tried not to look anywhere except the next rope rung on the ladder to be sure I didn't make a slip. I remember thinking, "Don't look down. Total concentration."

I was near the last rung now, and I knew the tricky part would be getting over the back deck of the Chinook with arms too tired to move reliably. The aircraft was higher in altitude now and bouncing violently in turbulence. I was determined not to get that far only to fall while climbing onto that aft deck. My left hand was clamped solidly in a death grip on the last rung of the rope ladder as it ran over the back deck of the Chinook. I reached into the aircraft with my right hand. I felt a clasp on my arm by my platoon sergeant. I gripped his wrist, and in that one final moment, I was in the Chinook.

It was my second-closest brush with death in Vietnam. Or maybe tied for first.

God was watching out for me.

We landed in An Khe, and I waited outside the aircraft for the pilot. I knew that what I was about to do was not smart and that it would land me in all kinds of trouble. The crew chief and the copilot saw me and must have warned him, because he never left the aircraft. I waited, and I could see glimpses of him in there, but he avoided eye contact. My platoon was loaded in the trucks close by, and all of them were watching.

Wisely, I let it go.

THE ENTREPRENEUR, DECEMBER 1965

IT WAS FADED BLACK and very old. The pickup truck rattled along a dike on the rice paddy.

Second squad spotted it at five hundred yards.

It was surprising to see a truck here in the rural highlands, where locals rode bicycles, plodded along with oxen-drawn carts, and tended the rice paddies on foot. To see a pickup truck, even one so small and old, negotiating these narrow rice paddy dikes was astonishing.

I radioed Frank Vavrek, Third Platoon leader, whose unit along with mine had been assigned a week of defensive perimeter duty.

I said, "Look behind us. Are you seeing what I'm seeing?"

He said, "We've been watching it. What do you think he's up to?"

The base camp defenses were set out in a series of perimeters. The truck was coming from my rear. To my front, and across the last line of concertina wire, it was bare and open ground to the jungle beyond. It had been cleared with Agent Orange and appeared apocalyptic. Not one blade of grass grew. Nothing.

No one could cross that barren ground to my front toward Hon Cong Mountain by day without being seen. It looked like Death Valley. It stretched for hundreds of yards toward the edge of the dense jungle. Just like the outer defense of most other outpost bases in Vietnam, the sandbag bunkers and foxholes were placed to provide interlocking fields of fire. Any attack by daylight would have been suicidal. At night, artillery flares lit up this sterile wasteland in front of me. As the flares drifted downward, the motion caused mysterious shadows to materialize and disappear like ghosts on this otherworldly landscape.

To our rear were rice paddies. They were included within the base camp defenses so the local Vietnamese could continue to farm unmolested by the NVA.

The local population who lived in this zone were often infiltrated by the VC and could not be trusted. As a result, we not only had to watch carefully to our front, but we also had to be constantly aware of what was going on behind us. That's when the truck appeared, behind me and far away.

I watched carefully as it motored slowly, maneuvering from one rice paddy dike to the next. How it managed that I'll never know. The tops of the dikes were thin and tapered on each side. I had scouted them.

What kept the truck from sliding off? The driver had obviously done this before. Who was he? Why was he here? I watched through my binoculars, searching for any clue to help unravel this mystery.

The truck soon stopped, and it looked to me like poles, or something shaped like poles, protruding from the back. A small, slender Vietnamese man in black, pajama-like clothing exited the truck. He wore sandals and a conical pointed hat made of bamboo and palm leaves. I watched for any sign of a weapon.

Wait. I saw him reach for something in the back of the truck. A rifle? Grenade launcher? I did a double take. Was that a post-hole digger? He reached in the truck again and removed what appeared to be a sharpshooter shovel.

All morning long he dug holes. He was a hard worker. But what was he up to?

He walked to the back of his truck and removed something larger. A wheelbarrow and a bag of something. Was that cement?

Soon he was mixing his cement and setting poles in the ground. Why would he do that on a rice paddy dike? Was it some type of fortification? I had to know.

At 1800 hours and before the curfew on the locals, he drove away.

Early the next morning he was back with a load of materials. It appeared to be two-by-fours and metal roofing. He worked alone and fast. Soon the support structure and roofing were in place on this open-air, pavilion-like structure. He left at noon but at 1300 hours he was back again and finished it. This guy was good.

By now I was glad he was there. It gave my platoon sergeant and me something to watch besides the tree line on the other side of the death zone in front of us.

I radioed Vavrek. "Come on over and watch this guy with me. Let's figure out what he is up to."

In no time Frank showed up. We watched the man through our binoculars, and I said, "He's reaching in the back of his truck again, Frank. Can you see what he is doing?"

Frank said, "It looks to me like a Samsonite folding card table and folding chairs."

"What on earth is he going to do with that?" I blurted.

Finally he nailed a piece of flat plywood to one of the poles facing our direction and began painting it. Through our binoculars we made out the letters as they were painted, one at a time.

RESTAURANT

They say in retail, it is all about location. But seriously, here in a war, in this remote location, on a dike in the middle of a rice paddy?

Was this guy nuts? Was he a Viet Cong sympathizer? A spy?

Or was he just some poor guy caught in the middle of all this insanity and trying to earn money to feed his family?

I hoped for the latter.

I said, "Frank. I've watched this fellow for two days. Maybe he's a VC and here to spy on us, but I doubt it. I like his spirit. Maybe he's got a family. Children. Maybe he can't make it with only this poor rice paddy."

Frank, who always had clever comments, chimed in, "Or maybe he believes in capitalism and is trying his best to get ahead, and has few options."

Either way, we knew we were going there. It took a few minutes to let our platoons know what we were doing and to cover us if it turned out to be some kind of trap. But I was confident it was not.

Vavrek and I, with our M16s, quickly covered the distance using the narrow dikes, never even getting our feet wet in the rice paddies. As we approached the pajama-clad man, he bowed slightly, and we did the same.

He motioned us to the card table and folding chairs in the shade on the dirt floor of his new open-air restaurant. With the language barrier, there was no real way to communicate with him. It was all head-nodding, gesturing, smiling, and sign language.

We laid our M16s on the table within arm's reach. He gave us two sheets of white notebook paper on which he had written in marker the word Menu. Below that, in pen and ink, there were only two menu items he had written in English—Water Buffalo Steak Sandwich and Bottled Coke, along with the price, 30 piastre.[5]

The menu selection was limited, but it made our entree and drink choices easy, and we placed our order. We had been eating C-rations for a while, and anything different sounded splendid. I was glad the Coke was bottled. That probably made it safer to drink and harder for someone to poison us. I had never heard of anyone eating water buffalo, but on a second thought, I could see why they would: I had seen no cattle in this country—only water buffalo, and of course the oxen for the carts.

It was a treat just to sit on chairs. They were hard metal folding chairs with no cushioning, but it sure beat the foxholes. We had been on a long-range patrol for a couple of days prior to pulling perimeter-defense duty. A chair, even a hard metal chair, was a comfort we weren't used to.

The proud proprietor bowed again, smiled, and said something we did not understand. He pulled out another piece of notebook paper and drew a clock, helping us to understand that he would return in one hour with our food. He motioned for us to remain and drove away in his truck.

Vavrek said, "Do you suppose he's off to kill a water buffalo?" I said, "If he does, it's going to be a very large sandwich."

Frank added, "I'm sure his meat-processing plant is FDA inspected and his workers are highly trained in restaurant sanitization procedures."

We chuckled and wondered when the restaurant health inspectors would be here. We tried to imagine the sandwiches being assembled. Each scenario we concocted seemed more bizarre and alarming. We decided it was best if we just didn't think about it.

In less than an hour, we saw the little black truck in the distance headed in our direction. Our waiter, maître d', chef, owner, and water buffalo hunter returned as promised, even a few minutes ahead of time.

He opened the truck door, smiling broadly. With a flourish he carried the sandwiches and Cokes to our table and proudly laid them out in front of us,

---

5. After the war, Vietnam changed their currency to the dong.

each wrapped in waxed paper. He had a bottle opener, and with great fanfare, he popped the caps off the Coke bottles right in front of us. It was reassuring.

The aroma was much like an American steak sandwich on hot toasted homemade bread. The bread was long and shaped like a deli bun, but wider and with a light-brown crust like French bread.

It smelled wonderful, and I couldn't wait to bite into it. When I did, I tasted something savory for the first time in months. But that was all I could do. It took many tries for my teeth to even tear through the bread. I had never encountered bread that tough. Then when I finally broke through the tough bread to the rich-tasting meat, it was the end of the road. As hungry as I was for something more than C-rations, and as hard as I strained and tried, my teeth would go no farther. Not only could I not bite through the water buffalo meat, my teeth never even dented it.

I looked at Vavrek, and he was struggling, too. We burst out laughing at the total absurdity of the situation, the war, the rice paddy, the choosing of this retail location, the comical and futile way we were struggling to chew, and the whole notion of a restaurant in this out-of-the-way place.

The proprietor became concerned, and I somehow let him know that all was okay and not to worry.

But every time I looked at Vavrek, we exploded in laughter. We had yet to successfully chew one bite of these delicious-tasting sandwiches.

I said to Frank, "I've got an idea." I placed our chairs facing each other. I braced my feet against the front of his seat and said, "Bite into your sandwich as hard as you can. I am going to pull on the end of your sandwich with both hands. You brace your feet against my chair, and you do the same for me."

It took two airborne Rangers pulling full strength with both hands for our teeth to finally break through that first bite of water buffalo steak.

It was a small victory, but the real battle was yet to come. Now we had to chew it. This we found to be really difficult, partially because we were laughing so hard. The world had never seen meat that tough.

Even the proprietor was beginning to loosen up and laugh with us. Or was he laughing at us?

I checked my watch. It was 1645 hours. That's when we started chewing. It took 15 full minutes to chew one bite. The more we chewed the funnier it got. All three of us.

It was a poignant moment on many levels as cultures formed worlds apart connected for a moment on a dike in a rice paddy in the middle of nowhere. Vietnamese, American, rice farmer, soldiers, platoon leaders, and restaurant entrepreneur. It occurred to me that maybe we are all not so different.

It was 1700 hours, and we had managed to chew only one bite each. It was time for us to leave. We paid our new friend, said our goodbyes, smiled, laughed, shook hands and then we were gone, back across the rice paddies to our platoons.

Frank and I eased back into my foxhole. My platoon sergeant was there. The three of us watched the faded truck as it skillfully negotiated the many narrow dikes until it was nearly gone in the distance.

My platoon sergeant asked, "Where is he going?"

Vavrek and I both smiled and gave the same answer, "To hunt another water buffalo."

It's been years since the end of that strange war, and to this day, the memories of it remind me of how much we have in America to be thankful for. In Vietnam I saw men like the entrepreneur, civilians caught up in that war, people who struggled to overcome what so many of us in this country have never experienced. And quite often, that memory makes whatever problems I'm dealing with at the time seem trivial.

It's been fifty-eight years since I sat in that foxhole and watched the building of a little makeshift restaurant in the most unlikely of places, and occasionally I've wondered what became of the owner. He probably runs one of those trendy new restaurants in Saigon (now known as Ho Chi Minh City). I hope so. I'll bet he's serving fresh fish, mollusk, and crab to wealthy patrons—or maybe he has a deli in Hanoi with a delectable menu of rice cakes, pork pastries, and ground beef wrapped in la lot leaves.

Or is he still in An Khe serving water buffalo steak sandwiches? If so, I hope he has found a higher-quality supply of water buffalo, with meat so tender it melts in the mouth.

At the very least, I imagine him still tending his rice paddies in the Central Highlands. But always I wonder—was he a VC? A spy? Or an entrepreneur?

I guess I'll never know.

N
W
E
S

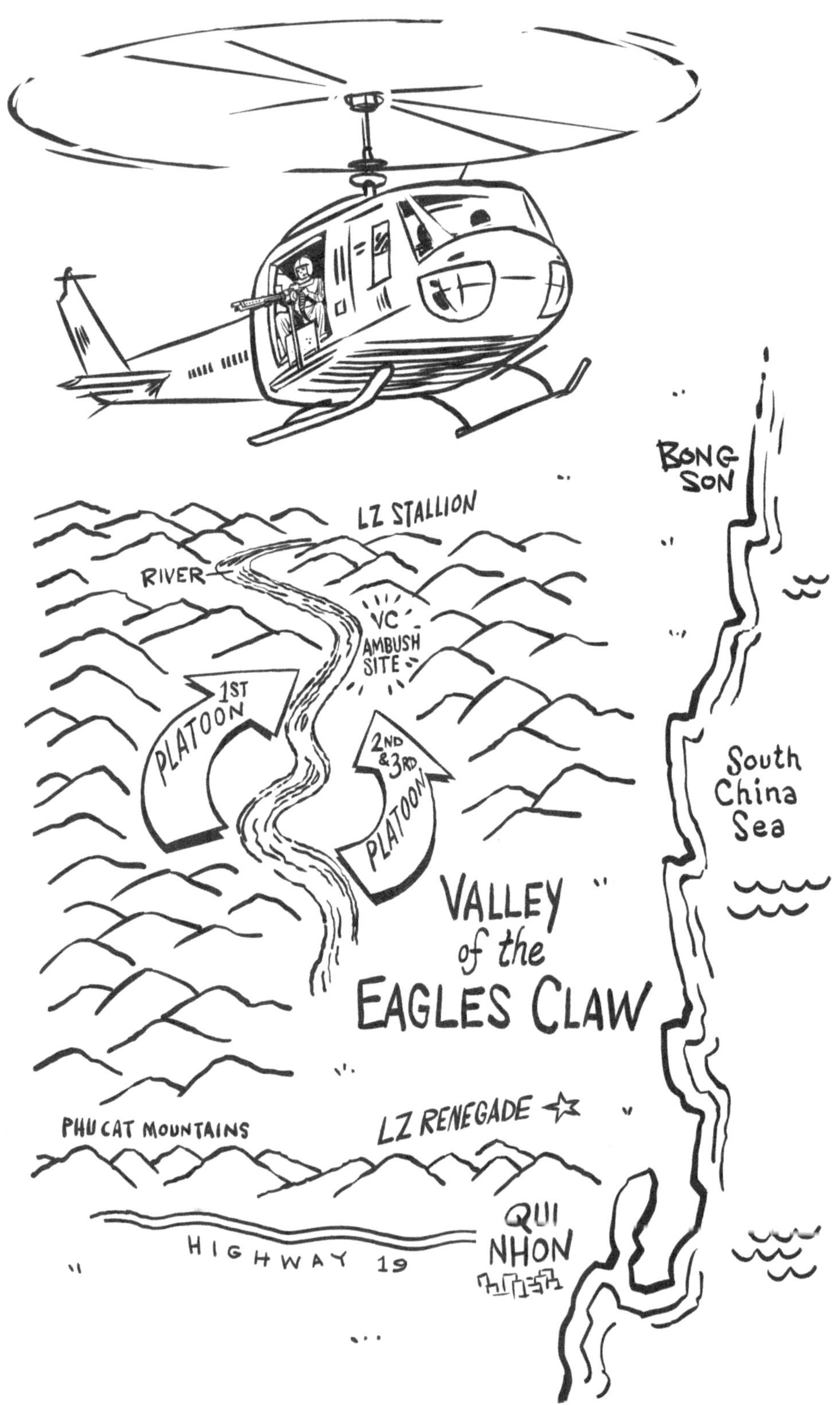
BONG SON
LZ STALLION
RIVER
VC AMBUSH SITE
1ST PLATOON
2ND & 3RD PLATOON
South China Sea
VALLEY of the EAGLES CLAW
LZ RENEGADE
PHU CAT MOUNTAINS
QUI NHON
HIGHWAY 19

THE CLAW OF THE EAGLE, FEBRUARY 19–20, 1966

FROM THE OPEN DOOR of the Huey, I saw the beautiful Phu Cat Mountains rising out of the South China Sea, its slopes lush and green in the morning light. The trees and canopy of the jungle were tall and thick while the mountain peaks were grassy, like the balds of the Smoky Mountains back in East Tennessee.

I listened to the rhythmic thump-thump of the rotor blades and felt the rush of wind on my face and the heat of the hydraulic system on my back. Below me was the long jungle ridge where, back in October, I had led my platoon on ambush patrols. We had fought in two skirmishes down there and had come out on top.

Now, five months later, we were back, flying over this same place. I felt the butterflies in my stomach just like the first time when I was here.

Ahead, in these mist-covered mountains and just south of Bong Son, was a secluded valley, an NVA stronghold, which according to S2 (intelligence) was occupied by a battalion of NVA. It was surrounded by the Phu Cat Mountains to the south, another high mountain range to the north, and the South China Sea to the east. It was known as the Eagle's Claw.

The previous evening, I had given the warning order to my squad leaders. I had told them the operation would commence at 0900 hours and for each of their men to carry four grenades, five hundred rounds of ammo, two days' worth of C-rations, a poncho, a poncho liner, shovel, two canteens, purification tablets, and a bayonet. I wanted to be sure we had plenty of ammo because we would be on our own, far from reinforcement by the rest of the battalion.

An hour later I had called the entire platoon together for the operations order. I told them we would be going into a small clearing, in a valley north of Qui Nhon and the Phu Cat Mountains and south of Bong Son.

Charlie Company would be operating alone. The rest of our battalion would air assault into the high mountaintops far to the north of us on LZ Stallion. We would be out of range of battalion reinforcement or mortar support. However, our Charlie Company weapons platoon would airlift to a lower ridgeline below LZ Stallion and would be in range to provide 81-mm mortar support if needed. I told them our mission was to search out and destroy any enemy units moving away from the battalion on LZ Stallion and down the long draws into the Eagle's Claw. I let them know the operation would commence at 0900 hours.

We had left base camp earlier that morning with Charlie Company loaded on 24 UH-1Ds, and now we were getting close. I felt the acceleration as we descended rapidly and swept in low, down the Phu Cat ridgelines and even lower, at treetop level, flying full tilt through a small finger valley. The object of flying so low and fast was to hopefully not allow the NVA to see where we landed or have time to draw a bead on us until after we were past them. For that same reason, we were going in with no artillery, mortar, or air force prep so as not to alert the enemy to our exact location.

We flew fast and aggressively, riding the terrain at treetop level like a roller coaster. The landing was swift in the small clearing with no resistance. I hastily moved my platoon into a defensive position on the east side of the LZ.

Once the remainder of the company was in place and the LZ was secure, I moved my platoon north as the lead platoon in Charlie Company through heavy jungle into the valley of the Eagle's Claw. Meanwhile, the rest of the battalion landed on LZ Stallion high atop the mountain range far to our north. They too met no resistance.

Throughout the day, I advanced First Platoon up the valley. The jungle was thick, and we came to a trail. I put security teams out about two hundred meters in each direction on the trail before moving my platoon across. Each of the other platoons following us did the same, and one of the security teams captured a lone NVA on the trail.

Captain William Mozey, our new company commander, called for a chopper. Captain Mozey had come down from battalion to take control of Charlie Company and replace Captain Smith, who had been our company commander since we left Fort Benning.

We wanted to get the prisoner back to battalion headquarters as soon as possible. To that end, we located a little clearing for the chopper to land and sent the prisoner to LZ Stallion. Through the battalion interpreter, he told us there were three NVA machine gun positions where the three valleys converged off LZ Stallion and into the main valley of the Eagle's Claw five miles to our front. With that in mind, we continued to advance cautiously through the jungle in that direction. As night approached, we settled into a small clearing and set up a defensive perimeter.

Meanwhile, Bravo Company was attacked up on LZ Stallion. Bravo repelled the attack in the dark and suffered only one American killed. But down where we were, it was all quiet—maybe too quiet.

The moon and stars had vanished, blacked out by heavy fog. I felt a strangeness here. There was an eerie silence, and it was darker than coal.

I was called to the Charlie Company command post. My order was to pick six of my best men and lead a night patrol through the jungle for a mile to a stream, locate a rope bridge built by the North Vietnamese, set up an ambush, and return by daybreak. There wasn't much time. I located the coordinates on the map, plotted our route, and chose an M60 machine gunner, an ammo bearer, and four riflemen to accompany me.

We left our defensive perimeter in pitch-black darkness with me as patrol leader, point man, and compass man combined. I had to assume the NVA could be anywhere, and I hated the thought of stumbling into them in the dark and in terrain I was unfamiliar with.

The jungle was thick, but we moved silently and deliberately. I was determined not to use a trail or even a creek bed, where the walking would be easier and faster but where we might be ambushed.

By 0100 we had covered most of the distance to the stream. I knew the chances were great that the NVA might be using the bridge or at least have guards there, and I advanced the patrol carefully in the darkness. As we neared the objective, I got a call from our company commander to abort. Higher-ups no longer wanted patrols out. They needed the entire company ready to move out at daybreak. I turned the patrol back toward Charlie Company and the remainder of my platoon. We moved silently through the jungle and by 0300 we were safely back inside friendly lines.

At first light of day, Charlie Company was ordered to air assault farther up the valley and intercept and destroy any enemy we encountered who might be moving down the narrow draws away from our battalion on LZ Stallion.

The choppers were on the way to pick us up, and we would land in the claw of the eagle, where three smaller valleys fed out into the main valley. On a map, the three smaller valleys looked like the eagle's front talons, and the main valley looked like the foot.

A flight of Hueys swooped in and lifted us out. Minutes later, we landed in the open fields of the main valley about five hundred meters from where the smaller middle valley emerged. I spread my squads out widely on-line since we were exposed in open terrain. As I advanced my platoon, I couldn't help but think about the prisoner's report of the three enemy machine gun positions right where we were headed.

I moved one squad at a time across the open ground. Once one squad was safely behind a dike on a rice paddy, I moved the next squad across. We leap-frogged each other, with some of my squads always in a position to provide fire support. At any moment I expected machine gun fire from the hills in front of us on each side of where the middle valley emerged. To me, it seemed odd that we had come this far with no enemy resistance. My rifle squads were on high alert, and we all moved quickly.

Soon we were beside a clear and rather large mountain river tumbling and swirling out of the middle valley before us. Anyone who has ever lived near the Great Smokies or most anywhere in Southern Appalachia would recognize this scene: each side of the valley was steep and very heavily vegetated in every imaginable shade of green.

Our new company commander, Captain William Mozey, and his RTO (radio man) were following my platoon, which was in the lead. This was his first real combat mission leading a rifle company.

I had two point men way out in front as we moved, and I stayed in visual contact with them. As we entered this narrow valley, something didn't seem right. I felt as if something sinister was watching, and that whatever it was might be close. All my Ranger training was telling me something about this place.

The left side of the river facing upstream was lightly vegetated and flat, maybe a bit too open. It's where the North Vietnamese would expect us to advance. It would be easier walking there than on the other side of the river where it was rougher terrain, steeper, and more heavily vegetated. About four hundred yards upstream, the river came pouring around a sharp bend from the left and then straight toward us.

From this bend upstream, a well-placed machine gun could easily cover all the flat and open ground on our side of the river where my platoon in the lead and the rest of Charlie Company were about to proceed—a perfect kill zone.

I also saw good hiding cover on the other side of the river where NVA infantry would have us in deadly crossfire when combined with fire from the bend in the river upstream. They would have the river as a barrier of protection and a second killing zone in case we tried to fight our way out of the ambush.

The voice in me was clear, and the voice said "trap." And I knew: if I were a North Vietnamese commander, this is where I would ambush Americans.

I stopped my platoon and radioed Captain Mozey to come forward so I could show him what I was seeing. I told him, "Sir, this is a likely spot to be ambushed if we advance up this side of the river. If they are dug in along the tree line over there across the river, and if there is a machine gun up there on the river bend , we'll be in a crossfire with no cover and no way to fight our way out because of the river. But if we were to move a platoon to the other side of the river, where the walking is tougher, we'll catch them by surprise. They'll never expect us to approach from over there."

Then I suggested, "Let me take my platoon forward, but up on higher ground on this side of the river where there is good cover. They won't even be able to see us up there, and if it is an ambush, we'll have them in a crossfire."

He quickly surveyed the situation and agreed. He ordered Vavrek's Third Platoon to the far side of the river, followed by Jon Williams's Second Platoon.

Captain Mozey and his RTO stayed just behind me as I moved my rifle squads a hundred yards up the moderately steep jungle mountainside on river-left looking upstream. Once I gained what I felt was the right elevation advantage, I radioed Vavrek, now on the far side of the river, to begin moving his platoon forward upstream, keeping up with my platoon but staying at least two hundred meters downstream from us as we advanced. That way, if the North Vietnamese were set up in ambush along the other side of the river, we would have them in a crossfire.

The entire company moved forward in unison, and if there was an ambush, not a single American would be visible to the enemy or in the kill zone. This is the kind of warfare Rangers were trained for—the kind I was most comfortable with.

It was now 1330, and our company was pushing forward cautiously and alertly up both sides of the river at two different elevations, and not on the easy, more open ground on river-left, where the North Vietnamese would expect us. We moved silently.

Suddenly there were grenade explosions and heavy automatic weapons fire from the other side of the river. Vavrek's two point men had stumbled onto the left flank of a North Vietnamese unit set up along the river in ambush. One of the point men was wounded, but the rest of Vavrek's lead squad pushed forward to assist the wounded soldier and engage the NVA at very close range.

While all of this was happening, the NVA hadn't seen my platoon higher up in the heavy foliage and across from them. I led First Platoon quickly forward through some thick trees to an area with cover and a good view of the tree line on the far side of the river. My men opened fire all at once. The left flank of the NVA ambush was caught in a crossfire between my platoon and Vavrek's. The noise in this narrow valley was now deafening, like gunfire in a tunnel.

At this point the left flank of the NVA realized they were in trouble. The ones nearest to Vavrek's lead squad started falling back upstream. But the rest of their unit was holding fast all up and down the tree line on the far riverbank.

I knew the NVA were in a fix, because for them to fire at Vavrek's men meant firing across their own left flank and killing their own people. The only platoon they could fire at was mine, and we began receiving fire from across the river as the NVA slowly realized where we were.

But we had the advantage in elevation, and it was difficult for them to see us in the heavy foliage.

I hollered at my weapons squad leader and led his squad forward with two M60 machine guns to a better position to fire directly into the tree line as well as cover the bend of the river upstream, where I was still convinced there would be an NVA machine gun.

Meanwhile, my three rifle squads were wreaking havoc on the NVA below and across from us. Enemy snipers were in the trees and dug in below the trees along four hundred yards of the far riverbank. But they were confused, because we weren't in the killing zone where they had expected us and wanted us to be. By this time, Vavrek had moved another squad forward and was tearing up their left flank.

I ran back to first squad. They were emptying one clip after another into the long stretch of trees on the far side of the river below and across from us. I knew full well that we had the upper hand and the NVA could not stay there for long and fight.

The North Vietnamese snipers and infantry in the tree line closest to Vavrek's platoon were continuing to fall back to our left, which was upriver. I could see them jumping from their concealed positions high in the trees and retreating farther upriver while carrying their wounded and dead with them.

Captain Mozey was on the radio and calling for mortar fire. He was directing it all up and down the riverbank across from us. I was impressed that he got it zeroed in that quickly and accurately. Simultaneously he was trying to coordinate an air strike, but we were told the valley was too tight and narrow for the fighter jets. Somehow he managed to find two navy World War II prop fighter planes that were holding over the battle zone. They were Douglas A-1 Skyraiders.

They came roaring in from the mouth of the valley to get a fix on the target. Vavrek and I both threw yellow smoke grenades to let them know where the lead friendlies were on each side of the river. They made a pass up the river to my left, pulled out, circled around, and on the second pass they came in low, showering the NVA with machine gun fire.

I was amazed at how low these two pilots could fly in such a tight little valley. On the second pass, one of the pilots was at eye level with me and couldn't have been more than twenty yards away. Each of them unleashed eight hundred 20 mm rounds on the four hundred yards of tree line on the far riverbank. I felt like I was watching a World War II movie close-up and in 3D, though I realized full well that I was one of the actors.

Suddenly it looked like nearly every bush and shrub on the far side of the river was moving. I did a double take. An entire platoon of NVA, each with large bushes, branches, and leaves tied and strapped to their bodies and helmets as camouflage, was abandoning ambush positions, popping out of concealed ground holes, jumping out of trees, and retreating as fast as they could up the valley toward the bend in the river to my left. At this point all three of my rifle squads, with their M16s, were emptying one clip after another.

In full retreat, the enemy fire withered, and just like that, it was over. But the adrenalin lasted longer.

We got a call from up on LZ Stallion. The colonel knew we were out here in no man's land on our own and beyond battalion support. Our order was, "Hold up where you are. We hope you haven't engaged a force too large for you."

Captain Mozey assured them that Charlie Company was okay, that the NVA had been set up in ambush, and that we had outflanked them. He reported one wounded from Third Platoon.

The colonel notified us he had a medevac chopper on its way for Vavrek's wounded man and that he was sending a flight of Hueys in early the next morning for the rest of us. He wanted Charlie Company linked back up with the rest of the battalion since the NVA now knew our location and that we were only a rifle company.

My new company commander and I sat down on a rocky outcropping overlooking the river and battlefield. We smoked cigars with my squad leaders while Second and Third Platoons scouted the rest of the ambush area below. I liked my new commander. Once he had a full view of the situation, he moved quickly, took control, and acted decisively. Charlie Company was in good hands.

I led First Platoon back down the mountainside to the river and across, sending two rifle squads upstream about four hundred meters, one on each side, for security. Then the rest of us began "cleaning up the battlefield." We found fifteen discarded enemy rifles with abundant gear and ammunition.

With only one from Vavrek's platoon wounded and none from mine, we estimated eight NVA killed, based on visual sightings. By 1700 hours, we moved back to the main valley. At nightfall we formed a tight company defensive perimeter.

The evening breeze was warm. The moon and stars were back out, and we watched carefully from our defensive perimeter, paying close attention to the shadows dancing in the trees.

On a night like this with no patrols, there was time to think—to replay the events of that day and the other battles in Vietnam.

For the first time in my life, I fully contemplated the importance of decisions and the profound effect a decision in battle can have on not only the soldiers but also their families and even future generations. Such decisions, made in complex battlefield situations, are based on little more than a hunch, on intuition and experience.

I am eighty-two years old now as I write this story. For me, it's been nearly sixty years since Vietnam, but somehow, I recall the ambush patrols and firefights with clarity, like they were yesterday. And I remember the narrow valley in the Eagle's Claw, and the decision I made that day—and often I've wondered what it would have been like if Charlie Company had walked into the kill zone.

Charlie with a brown trout from the Little Tennessee River (the Little T) near the southern border of the Great Smoky Mountains National Park, 1969.

Fly rod world record blue marlin, 208 pounds, 16-pound test tippet, Cape Verde Islands, 1998. *Left to right:* Captain Trevor Cockle, mate Ronnie Fields, angler Charlie Tombras, mate Randy Baker.

Charlie—headed into Cape Verde with a potential blue marlin fly rod world record, 1998.

Charlie with a 13-pound rainbow, Kvichak River, southwest Alaska, 1998.

Charlie's eighth fly rod world record, 130-pound Pacific Sailfish, Guanacaste Province, Costa Rica, 2002.

Dooley and Charlie in Miami, 2009.

Carol, Charlie's beloved wife of 34 years, who passed away in 2011.

Dooley and Kyri on the North Fork at the ranch, 2010.

The author with a nice brown trout in Argentina, 2015

Charlie on horseback at a backcountry lake, high in the Mount Zirkel Wilderness near the ranch, 2016.

Left to Right: Brenda's grandson, John Charles Trotter, Brenda Tombras, Brenda's son, John Trotter at the ranch, 2016.

The author with a 25-pound brown trout in Argentina, 2017.

The author with his two setters, Roamer and Whitey, 2018.

At their home in Knoxville, Brenda and Charlie, 2019.

Brenda and Charlie. A morning ride on the lower ranch, 2020.

Brenda and Charlie at the ranch, 2022.

PART FOUR

Homecoming

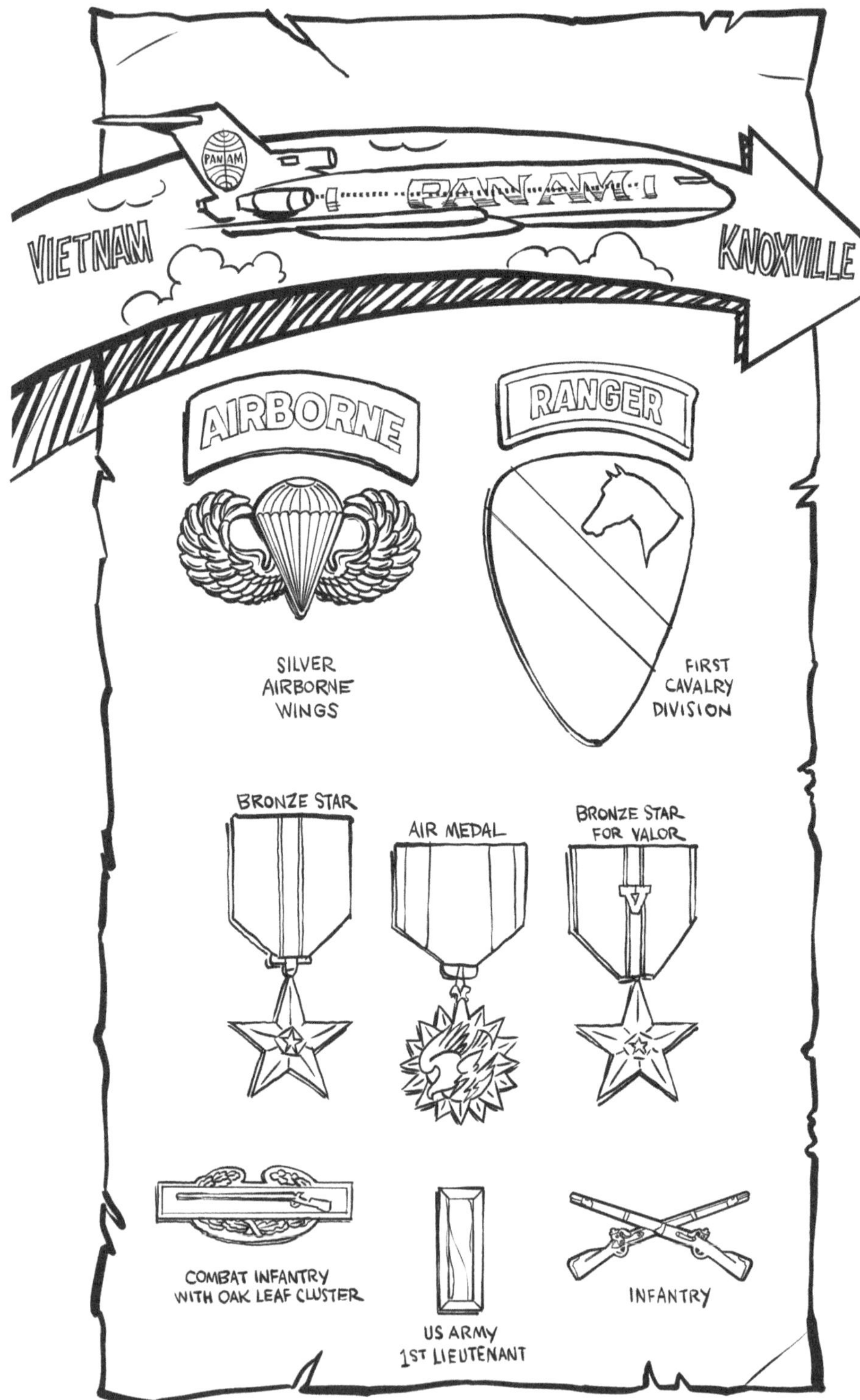

PAN AM
PAN AM
VIETNAM
KNOXVILLE
AIRBORNE
RANGER
SILVER
AIRBORNE
WINGS
FIRST
CAVALRY
DIVISION
BRONZE STAR
AIR MEDAL
BRONZE STAR
FOR VALOR
V
COMBAT INFANTRY
WITH OAK LEAF CLUSTER
US ARMY
1ST LIEUTENANT
INFANTRY

STARTING OVER, 1966–1973

MY TOUR OF DUTY was over, and I came home alone. I was twenty-four years old.

There were no Vietnam buddies returning to the States with me. No homecoming parades. No welcome home rallies. Nor were any expected.

Many of my friends said, "Thank you." A few asked where Vietnam was. One day I was in a war. Two days later, I was here.

It was wonderful to be home with family and those I hadn't seen since college, but beyond that circle, conversations often felt awkward. There were polite questions about the war, what I thought of it, what I saw, and what it was like. But I never felt that I was communicating in a way that conveyed the enormity of it or did justice to those brave men I served with.

In Vietnam, conversations were serious. The outcomes could be life or death. Here at home, at a gala or cocktail party, the conversations seemed less important and occasionally superficial. People meant well. They were just trying to be friendly and make small-talk, and I appreciated that, but I found it hard to focus. My mind would drift to a place far away.

I was glad they hadn't seen what I had seen and been where I had been. And because they hadn't, they couldn't grasp the magnitude. Or was it me? Perhaps I couldn't find the words to convey how fortunate we all were to live in this wonderful place. For some, it was a given. And I had trouble with that. I had forgotten that until you weathered a storm, and weathered it personally, it could be difficult to fully grasp what is real and important, lasting and meaningful—the things that really matter.

But there was something more that I struggled with—the language of society. It was a language I used to speak fluently, but one in which I was now two years lacking. Often I was inept, and for the first time in a while, I felt inadequate.

To be clear, I had no PTSD, no emotional problems. I was at ease with my family, close friends, and especially anyone who had served in the military. But beyond that, I felt slightly out of place—in a place where two years earlier it had felt so right. I was fully accepted, and yet there was this nagging sense of unbelonging. A feeling of isolation—that I was better at what I had done than what I was doing. Often, I felt like a stranger that knew too much, yet had no means to communicate effectively, or anyone capable of hearing or understanding.

I was here, but not completely. Part of me was still there.

As I drove to work each morning, I found myself scanning every ridgetop, creek, and tree line as places to defend, to avoid, or to have a strategic advantage in a firefight. It was like I was not yet totally here.

Somehow the war had changed me. I was no longer who I used to be, but unsure who I was or would become.

I wondered if returning home so quickly had made it unnecessarily difficult for me to make sense of the two worlds I had lived in. Or was it that I came home alone, not with my army unit, like the soldiers in World War II? There were no war buddies coming home with me who spoke my language.

There was never any hate toward me. No one spoke a bad word. But to me, it felt very much like polite indifference.

It bothered me.

One day I was an infantry rifle platoon leader in a war. Two days later I was working at the Tombras Group, the company my dad had started in 1946, and people in the advertising business were very polite, but indifferent.

Finally I figured out what the problem was. It was me.

Instead of being a platoon leader with combat experience and looked up to by many, I was a new guy in the advertising business with no ad agency experience, no media experience, and no client experience. I was looked up to by absolutely no one. It was going to take some adjustment. I was starting over, and at the bottom. I had forgotten how humbling it was to start at the bottom, no matter the endeavor—to prove myself all over again. And it was especially difficult for me, transitioning from running a rifle platoon, where I excelled and where people looked up to me, to entering a business where I had absolutely no experience or credibility.

Once I grasped that, I accepted the indifference.

Whether you are starting or starting over, you had better have a goal. I wrote mine down and carried it in my wallet for forty years. My goal was to grow our little ten-person local ad agency into the largest ad agency in Tennessee.

I thought it was an aggressive goal considering how small, unknown, and under-capitalized our little company was. In retrospect, it wasn't nearly bold enough. I accomplished it and more—not through acquisitions, private equity investment, or debt. I did it the old-fashioned way, through hard work and taking market share.

During my first five years of working at the Tombras Group, I seldom took a vacation, a sick day, or a day off. Mostly I worked ten- to twelve-hour days, including weekends. Many of my friends who had not served time in the military had a two-year head start in business, and I was trying to make up for lost time. But even beyond that, there was the learning curve of the advertising agency business itself, and I tried to accelerate that curve.

It was during this time that I met my wife-to-be, Elizabeth Collins. She was an attractive girl who had grown up in Knoxville and graduated from the University of Tennessee. We married and had one son, Charlie. The marriage lasted only a few years and we went our separate ways. Perhaps I was too focused on my work, or too young and immature in managing and fostering a relationship. I felt an overwhelming need to prove myself all over again in this new life beyond the military and saw little progress in that regard. I was not growing, personally or in business, and it caused frustration that seeped into my marriage.

Meanwhile my love of fly fishing, boat racing, and exploring the outdoors had not faded—it never did—but out of necessity, it took a back seat to business. I realized I had a better chance of one day enjoying all of that if I became financially secure, and the progress I was making in business seemed far too slow. I began to wonder if there was a better way and where this life was taking me. There were dreams that were unfulfilled in business, great rivers to be explored in fly fishing, and I wondered if the path I was on would ever get me there. I felt that I had so much more to accomplish without any exact sense of how that might unfold or the adventures that awaited.

By the mid-seventies that all began to change, and my dreams began to take shape. Our business at the Tombras Group[6] began to grow, and little by little, I became more like my old self again.

Slowly and surely, I began to set aside time away from business, reuniting with the fly rod.

6. The author's career in advertising is recounted in the epilogue, "The Story of the Tombras Group," written by Dr. Warren Dockter, president of the East Tennessee Historical Society.

N
W
E
S

Tierra del Fuego
STRAIT of MAGELLAN
RIO GRANDE
RIO GRANDE RIVER
CHILE
ARGENTINA
Atlantic Ocean
Pacific Ocean

THE END OF THE WORLD, 1975

THE LARGEST BROWN TROUT on planet earth were rumored to live in the Rio Grande River[7] in Tierra del Fuego, Argentina. Naturally, we went there, my dad and me.

That was forty-six years ago, before the advent of travel agents who arranged fishing trips and before there were fishing lodges in that part of the world. We flew to Buenos Aires, where I had hired a guide named Laddie Buchannan. The three of us met up at an Argentine coffee shop. The next morning we flew to Rio Grande in Tierra del Fuego, which is as far south as you can go in the Western Hemisphere. Well, not quite: the next stop is the Antarctic.

Laddie had fished Tierra del Fuego before. In fact, a decade earlier he had lived there and guided a fellow named Joe Brooks, who was one of the first fly fishing travel writers of significance in the world. Joe had written a book, and one of the chapters in that book was the account of his expedition down the Rio Grande River in Tierra del Fuego with Laddie. I must have read that chapter twenty times. I had never seen brown trout as large as the ones in those photos.

I contacted Joe, and he gave me Laddie's phone number in Buenos Aires. Joe, who was a great and well-respected fly fisherman, passed away shortly

7. Not to be confused with a river of the same name in the United States.

after that phone call in 1972, but I kept Laddie's phone number, and the following year I called him in Buenos Aires and asked if he would go to Tierra del Fuego with my dad and me.

Having lived in Tierra del Fuego, Laddie knew the estancia owners along the river. The plan was for Laddie to guide us for four days and introduce us to the estancia managers. Then Laddie would fly back to Buenos Aires, at which point Dad and I would be on our own.

The estancias were huge and seemed to stretch on forever. Back then, they were primarily sheep ranches. Most were twenty or thirty miles from one side to the other. It was big, desolate country.

We saw almost no trees on this vast and barren landscape. The few wild animals that existed in this inhospitable climate were called guanacos. They were very inquisitive wild herd animals and resembled a cross between a camel and a llama.

There were Magellan geese by the thousands, foxes, and an occasional South American condor that would sail off one of the distant snow-capped mountain peaks separating the Argentine part of the island from Chile. Other than that, as far as we could see, there was nothing. No people. No trees. And no houses, except for the small cluster at each estancia headquarters.

The wind was insane and ruined our car doors on both sides. We had to be careful to always head the car into the wind before we parked it on the prairie. We didn't one day, and the wind nearly ripped the doors off the hinges. The doors never shut completely after that.

Conditions for fishing were tough. The wind was so strong the waves muddied up the river. As a result, we alternated days, one day fishing the main Rio Grande and the next day driving all the way to the mountains near Chile to fish lakes.

It happened on one of those mornings when we were driving west to fish the mountain lakes near Chile: As we rambled across prairie, we saw a truck far in the distance. It was noteworthy because all morning long we had seen no vehicles of any kind. In fact, it was only the second vehicle we had seen that week. Laddie was driving. I was in the passenger seat, and Dad was in the back. I could see the truck miles away with clouds of dust billowing behind it in the wind.

As we approached from opposite directions, I saw the driver wave us to stop. It was an Argentine policeman. He got out of his truck and walked up to the window of our car on the driver's side.

He was from the town of Rio Grande. He spoke English. The very first thing he said was, "Laddie, where have you been for so long?"

Laddie said, "Up north, fishing for dorado."

They chatted for a while, and then the policeman said, "Be careful out here."

And Laddie asked, "Trouble?"

The policeman replied, "Laddie, have you got a gun?" Laddie said, "Well yes, I always have one."

The policeman asked, "Can I see it?"

Laddie said, "Yes, of course," and motioned for me to open the glove compartment, which I did and handed him the 44.

The policeman said, "Good. Can you shoot it accurately?" Laddie replied, "Yes, I shoot it very well."

The policeman said, "Excellent. Would you mind doing me a favor?"

Laddie said, "Yes, of course."

The policeman said, "Great. Let me show you this man."

He unrolled a poster with a picture of a bad-looking hombre. I could tell it was a wanted poster, but I couldn't make out the words in Spanish.

The policeman said, "Laddie, can you do me a favor? If you see this gentleman on the prairie, or one of the two-tracks—probably riding a horse—could you please stop your car about thirty feet from him? Don't get too close; he is probably armed. Take careful aim with your gun and shoot him." He then said, "When he falls, stay back away from him. Shoot him again, to be sure he's dead. I sure would appreciate it."

I could barely believe what I was hearing. And then Laddie said in the most matter-of-fact way, as if he were offering to help someone carry groceries in from the car, "You got it. Glad to help."

The policeman tipped his hat, got back in his truck, and continued across the prairie toward Rio Grande. Dad and I just looked at each other in astonishment. I said, "Laddie, what just happened?"

He said, "The guy they are looking for is a murderer from Chile, and he crosses the border up ahead and steals horses from the estancias and takes them back to Chile. The estancia owners want rid of him."

As if this empty and lonely land with its fierce cold wind was not sufficiently foreboding, this was a potential new challenge to contemplate. I felt as if I had traveled back in time a hundred years to Montana or Wyoming—only we were going to shoot even if the other guy didn't draw first.

Somewhere between the wanted poster and the desolation of this vast land, or both, I was convinced this was the most bizarre place I had ever been. The long and lonely dirt road to the lakes near Chile took on new meaning. A potential gunfight. I felt deputized, scanning the dusty road ahead and horizons on each side for some bandit on horseback.

Each day, we made the rugged and challenging drive from the little town of Rio Grande to the immense estancias and the river. We wasted valuable fishing time doing this.

The river meandered aimlessly all over the vast prairie, and sometimes it was a challenge just finding it in this treeless country with no visual landmarks. The road from town was a dirt road, and the ranches had only two-tracks or nothing at all. Nevertheless, it was pleasant having approximately seventy miles of river all to ourselves.

My dad seemed to take to this new river better than me. He caught an eighteen-pound brown trout the second evening while I was struggling to figure out this strange river. I had fished in Alaska and was convinced that these huge fish wanted big flies like the rainbows in Iliamna. Rio Grande trout do like large flies sometimes, but it's mostly in the late evening or just after dark when they do. And that was a problem for us, because the restaurant in town closed early, which meant missing dinner if we fished late.

Some nights, we missed the best fishing and enjoyed dinner in town. Other nights, we sacrificed dinner to fish in "prime time," which was the last hour leading up to dark and the first hour after dark.

When Laddie left and flew back to Buenos Aires, Dad and I would go to a ranch, get a key to the gate, and drive as close to the river as we dared. Sometimes we could not get within a mile of the river due to the lack of a two-track and the swampy bogs.

Our routine was to split up and meet back at the car for lunch, then maybe drive to a new area and fish until dark or close to it. We fished eight days on this desolate river and never saw another person. And then the next most bizarre thing happened.

It was well after twilight, almost completely dark. I was hurrying up the south bank of the river to find Dad.

Through the encroaching darkness I thought I spotted something, a glimpse of movement to my front. It startled me. I stopped out of caution, the thought of the horse thief from Chile still on my mind. Then it moved again. I couldn't make out what it was in the darkness, so I moved forward cautiously. As I got closer, I saw the shape of a man. He was lying on his back with his hand over his eye. As I inched forward, I realized he was wearing chest-high waders and a trout vest.

After all the time on this river without laying eyes on a living soul, I was amazed to see a human being. Much less to stumble across him lying there in the dark. It was profound.

I walked up to him, and I said, "Do you speak English?"

He was shocked out of his mind to see someone appear out of nowhere in this lonely place. When he regained composure, he replied, "Yes, I speak English."

I never expected him to speak English, and I said, "Is everything okay?"

He replied, "Well, no, actually. I was casting in this darn wind with my fly rod, and I'm afraid I've hooked myself in the eye."

"I have a flashlight in my pack," I said bravely. "Let me have a look at it."

I dug the flashlight from my pack and directed the light at his eye. What I saw was not good. The hook was buried through the center of his upper eye lid. It was a large streamer fly, and the hook was in all the way to the bend.

By then I was thinking, "How is the best way to help this poor man?"—which, of course, I was going to do. But I had to figure out how. I knew full well it was 9 p.m. and we were two or three miles across the prairie to the main dirt road and thirty miles or more to the small town of Rio Grande. And once I got him there, what then? It would be midnight, a strange country, a language barrier, and the chances of finding any medical help, much less the type he needed, was somewhere near zero. I was also mindful that my dad was somewhere upstream, waiting for me in the darkness, and he would soon be worried and searching.

While I was pondering these weighty thoughts, I looked up the river, and there, in the darkness, I saw movement once again—a faint figure of a man approaching, and he was close. Thinking of the wanted poster, I quickly said to the guy with the hook in his eye, "Are you by yourself?"

He said, "Yes."

I said, "Well, here comes someone else."

Out of the darkness walked another guy. He was as surprised to see us as we were to see him. The first thing he said was, "Do you speak English?"

"Yes," we replied in unison. Then he asked, "Is everything OK?"

I said, "Not really. This fellow has hooked himself in the eye with a large streamer fly, and I need to get him to Rio Grande and try to find a doctor."

I'll never forget his next words for as long as I live.

He said, "Oh, I'm an ophthalmologist." And then he added, "Let me have a look at it." I was stunned. My jaw dropped open as I stood there in disbelief, and I wondered if I was hallucinating. Was this even real?

Here in this barren land of Patagonian nothingness, a doctor had materialized. And not just any doctor: a doctor who spoke English. And not just any English-speaking doctor: an ophthalmologist.

Were these ghosts?

The doctor had a day pack and took it off. He asked me to hold the flashlight while he examined the other gentleman's eye. I held the light while he slowly peeled the eyelid back.

Next, he told the other fellow, "You're really lucky. The hook is through the eyelid, but it has only scratched your eye." He then said to me, "Do you have wire cutters?"

I said, "Yes," and pulled out my pliers that had wire cutters on the side.

He said, "Here, hold his eyelid open with one hand and shine the flashlight with the other hand, and I'll cut the hook, and we'll back it out." Which is exactly what we did. The whole procedure was done in less than a minute.

Then the doctor said to the other gentleman, "You had a very close call, and fortunately there won't be any permanent damage to your eye. It's a severe scratch, for sure. I have an antibiotic eyedrop." He dug into a pocket in his pack and said, "Here, take this. Use it four times a day until you get back to the States and see your doctor."

Then he dug around in his pack again and said, "Your eye is going to be real sore for three days. Here's a pain killer. Put two drops in your eye every four hours for the next three days and you'll be fine."

By this time, I felt as if I was in a twilight zone.

We discussed which direction each of us was headed. I told them I was going upstream to find my dad. The doctor said, "He must be far away, because I just fished this evening in the two long bends of the river just upstream from here and saw no one."

The gentleman with the damaged eye said his truck was way out on the dirt road to town and that he had a very long walk.

The doctor said he had driven his truck across the prairie, and it was near the river several miles downstream. So, there in the darkness of Tierra del Fuego, we wished each other well and walked off into the night, in three separate directions.

A half hour later and four long bends in the river upstream, I found my dad. He had been waiting for me in the darkness, not so patiently. As I walked up to him, I said, "Have you seen anyone today?"

He gave me that look I had seen a thousand times in my life—the one that said, "Have you lost your mind?" I'll never forget his reply. He said, "I haven't seen anyone all day. In fact, I haven't seen anyone all week. And by the way, what took you so long?"

I just said, "You wouldn't believe me if I told you." And together, we walked through the darkness of Patagonia to find our car.

GOIN' BACK TO MONTANA, 1977

LOOKING BACK ON THIS, *I can't believe I didn't carry the gun.*

I'd thought about it and decided against it. I was in a hurry, and besides, I had been here before and hadn't needed it. So I had stowed it in the glove compartment and locked the truck.

It was early fall, and the September wind brought a chill to the Centennial Valley. I packed my gear in the sagebrush at the end of this lonely dirt road. Overhead the wind came briskly off the Gravelly Range, and the little aspen leaves were quaking. Before me was a dark path through tall lodgepole pines where the slope dropped off sharply to the hidden lake, which was my destination.

I hiked the mountain trail alone. It was a place where others had been with me before. Friends and loved ones. Those had been the good years—back when Montana had first got its grip on me. In my mind, I could still hear their voices, but always from some place beyond where I could see—around the bend in the trail just out of sight, and sometimes in a faint echo off the rock scree high above. For a moment I felt they were close, and then I heard them no more.

The path to the hidden lake was remarkably quiet, and there was nothing but the silence, the musty smell of ancient trees, the darkness of the forest, and the solitude.

It had been a long drive from the Madison River up to the remote

Centennial Valley in the high country of Montana and the place I had left my truck where the trail began.

The last thirty miles of road was gravel and mud. It had rained, and the tires on my truck slipped and strained in the deepening ruts. Despite the road's condition, it was a drive I always looked forward to. Driving in Montana gives me time to reflect on fly fishing and life, and at the age of thirty-six, this high alpine valley had already given me a lot to remember and think about.

Those memories took me back in time to 1953. I had been a twelve-year-old boy in awe of his first trip west with his dad. Like most young boys, I had been on the trail of something that would lead me out of childhood.

We had fished the Ruby River. It was late in the day, and we had driven up its headwaters and all the way into the high Centennial Valley on a dirt two-track. Back then, there were no fences up here. The sun was low in the west behind us, and it cast a Hollywood glow on the prairie with snow-capped peaks all around. It was in that moment that I saw something that never left me.

A real-life high-country cattle drive. Cowboys off to bring large herds out of the mountains to the Madison Valley far below. Chuck wagons. Saddles laid out on the cold prairie ground. Bedrolls.

Horses hobbled nearby for the night.

Etched in my mind today are the campfires: steaks frying on open flames, the sweet and distinctive smell of mountain sage, the voices of the cowboys around the fires, the sound of their horses and cattle—and coyotes howling in the distance.

Montana is a land that can reach out and get a grip on you, and once it has you in its grasp, it never lets go. You can choose not to return to it, but it always returns to you. Even when you close your eyes and sleep, it's there, tugging at you gently.

Those were the thoughts that filled my head as I drove the wet and muddy road this morning.

Now the woods were dark. I walked the path alone and wondered if I had waited too late in the day to hike into this high alpine lake. Sometimes walking into deep woods is like going from sunlight into a dimly lit room. It takes time for your eyes to adjust, but today, somehow, the woods seemed even darker than usual.

The only sound now was my own footsteps on the trail and the occasional rhythmic movement of the float tube strapped to my backpack. I walked steadily and quietly. After a while the forest began to open, and finally sunlight came streaming through the trees. I began to feel better. Then the trail dropped more rapidly, the trees became sparse, and the hidden lake came into view.

It was turquoise like the rarest of western jewels and nestled into the Gravelly Range with the rest of the Madison Valley fading off into the distance far below. I stepped out of the forest into the sunshine and the green grass growing along the shore. The chill in the air was gone. I forgot about the lateness in the day and concentrated on getting into my waders, flippers, and float tube as quickly as possible. I stashed my backpack, took out my sandwich and stuffed it in the pocket of the float tube along with a box of flies, leaders, and tippets, and launched slowly into the lake.

In a float tube you launch by walking backward into the lake. Anyone who has ever tried to walk forward in flippers knows why. But even walking backward, you must move slowly and carefully so you don't trip over something in the water like a rock or a log.

Once you are waist-deep, you settle into the seat of the tube and propel yourself backward with the flippers. It's awkward at first, and important to look behind you every now and then to see where you are headed. To me, it is a very relaxing and pleasant way to fish. On backcountry lakes like this one, it's the only way to fish if you want to cover all the water.

The north side of the lake was moderately steep. Lodgepole pines grew to the water's edge. Past that, the breeze rippled the far end of the lake. But here in the shallows by the green grass, the wind was flat, and the water was still. The sunshine had warmed me, and the coolness of the water was refreshing against my waders.

I rigged three wet flies about three-and-a-half feet apart on a long leader. I love that moment when I am finally fly fishing. I was back and felt alive.

A decade earlier, my dad and I had fished here. Before that, we had fished this lake with others who were like family. I wished they were all here again.

I caught a brightly colored rainbow, then another, and another, holding each of them with wet hands for a moment before releasing. I was glad I had come back to this place. I was alone, and there was nothing else in my universe except the sparkle of this lake, the vastness of the forest, and the mountains of Montana all around.

The northwest side was always my favorite. The water was deep next to the shore. I eased my float tube along slowly, moving backward and close to the bank, trailing my three-fly rig on a full sink line, far behind and deep below the surface.

I came to the place where the sparse pine trees grew nearly to the lake but were shielded by shorter and stubby western willows that hung out slightly over the shore. I was silent in the water and easing along close to the bank—and that's when things got dicey.

Something erupted behind me, searing the silence. Mallards exploded upward in a mad flapping of wings, dousing my back with cold mountain water as the first mallard cleared my head by inches.

In that millisecond I thought it was me that had frightened the ducks. But before half the ducks cleared my head, I found that I was wrong. I wheeled around quickly to see where the ducks were coming from.

What I saw was so shocking that I froze, and for an instant, I could not move.

A grizzly was coming at me full speed, crashing through the short willow trees behind me and slightly above. The bear had already sprung and was in the air. I was so low to the water that the grizzly was above me with arms outstretched, claws spread, teeth showing, ears pinned back, headed down into a moving pile of frightened ducks—and me.

He came down hard, like a Greyhound bus, in an explosion of water next to my float tube. The tsunami nearly overturned me, but in the melee the float tube somehow stayed upright.

I was kicking my flippers as fast as I could underwater to move away, but momentum comes slowly in a float tube. The bear swam a couple feet to the shallow ledge on the shore and stood up on his back legs in the water. He was probably seven feet tall, and he was mad.

He was shuffling around on his two hind feet, clicking his teeth and trying to figure out what I was. I was hollering, waving my arms, and trying to get some distance.

He came down on all fours, pinned his ears back, and clicked his teeth some more, but instead of charging, he stood back up. I guessed he was trying to size up exactly what I was and why I was here in this place where only he should be.

For a moment he looked perplexed, not sure if I was a competitor and after his ducks, or if I was some other kind of threat—or if I might be the prey.

Low to the water in my float tube, I must have appeared nonhuman and I hoped even nonthreatening. I'm sure I looked unusual and weird enough that he was having trouble deciding what I was and whether I was his supper.

I took this as an opportunity to gain about twenty feet of separation. Maybe it was good that I was in the water and not on the shore. You can't run from a grizzly bear: that triggers an attack. Of course, you can't run with flippers on your feet anyway—not to mention the float tube.

If a grizzly charges and you don't have pepper spray or a weapon, the only defense is playing dead. You curl up tight, clasp your hands and fingers behind your neck, and tuck your knees up to protect your stomach and pray. Occasionally someone will survive such an attack, but I suspect the percentage is low. I wasn't calculating percentages. I was busy furiously kicking my flippers underwater to gain distance while trying desperately to figure out how to play dead in the water without drowning. It didn't seem like much of an option.

I knew that grizzly bears are great swimmers. Nevertheless, I felt that distance was my friend. My only game plan was to move quickly out into the lake and away from him. With my feet being underwater, maybe it would not appear to him like I was running, which would provoke a charge.

He came down on all fours again as if he might be thinking of coming into the water after me. I started waving my arms and hollering so he would know I was human. I knew that grizzly bears are not afraid of humans, but I didn't know what else to do. There was no other plan.

The hollering didn't scare him in the least, but it did make him stand up again on his two back legs again so he could once more try to figure out what I was. It gave me time to gain more distance.

This whole sequence repeated itself three or four times until I was a hundred yards out in the lake. Except for his first lunge from the bank through the air into the ducks and me, at no time did he ever get fully out of the water. Then, when I was out in mid-lake, he calmly came down on all fours, walked into the water down the shoreline, around the end of the willow trees, and ambled up on the bank a few feet. Then he stood up on his two back legs and tried to scent me again and get a better look.

Pretty soon he came down on all fours and paced back and forth along the bank, occasionally standing back up to look and scent. The whole time, he never took his eyes off me.

After twenty minutes he ambled up the hill through the sparse pines, stopping to look back at me and occasionally standing for a better look. Then he was gone.

At first, I was in shock. But now the full force of fear was kicking in. Drifting alone in the middle of this wilderness lake, I had time to think and to realize how close I had come to death.

It was now late in the day and the sun was almost touching the mountain on the west side of the lake. I didn't relish walking out in the dark. Grizzlies do most of their hunting in the dark. The woods I had hiked through were dark enough even in the sunlight. I knew it would be pitch-black in there if I waited too long.

On the other hand, I wanted to wait long enough for that bear to not be so focused on me. I moved over to the other side of the lake and fished the shore going back to where my pack was stashed.

Looking all around, I removed my waders and flippers, got my hiking boots on, and strapped my float tube to my pack. The sun had set with only a ray of light remaining on the highest of the nearby mountain peaks. It would soon be fully dark.

I started up the trail, and I thought of the gun, and how careless I was to have left it in the truck. I vowed never to repeat that mistake again.

I hiked the steepest part of the trail where the trees were sparse and there was still some twilight. I moved quickly but stopped every now and then to listen and watch behind me.

Then I came to the dark woods. It was difficult to see the trail now. I was moving fast, but my heart was beating even faster. I was thinking about the bear and the gun.

It was not a long hike, maybe a half mile or so. But in the Montana darkness, and for as long as I can remember throughout my life, it was the longest half mile I ever walked.

CAROL, 1978–2011

SHE WAS THE SUNRISE in my heart.

The first time I saw her, she was in the first grade, and I was in the fourth. I was nine years old. She was five.

Her name was Carol Smith. Her family owned a farm along Kingston Pike in what is now the Cedar Bluff area of West Knoxville. She and my sister, Nancy, were friends in the same class in Bearden Grammar School. Quite often, Nancy would invite Carol to our small summer cottage on the lake, and I taught the two of them to water-ski.

Our paths separated when I went off to McCallie Military Academy in Chattanooga and Carol stayed home at Bearden School and later at West High. When I returned from Chattanooga to attend the University of Tennessee, she was still in high school and four years younger than me, so we never dated, and I never saw her, even from afar.

Then one day she showed up as a freshman at the University of Tennessee. And wow, had she changed. She was first runner-up in the Miss UT Pageant, which was a very big deal back then, and she was featured on the UT calendar.

I was about to leave for the US Army and Fort Benning, Georgia, so as much as I wanted to, there was no time to ask her out. I saw her, we chatted once, and that was it.

I shipped off to Infantry Officer Basic Training, and it wasn't until years later that I saw her again. It was in the late 1970s, and there were a couple of nightclubs in West Knoxville popular with my age group. The music cranked up every night about 11 p.m. and went strong until 2 a.m. or so. That's when I saw her.

It looked as if she was trying to make her way over to me to say hello. The room was crowded, and every eye was on her. She was that stunning. When she smiled at me, it felt like something I had waited for my entire life—and it was in that moment that I knew my world would change forever.

I was thirty-six. She was thirty-three.

I called her the next morning. We went out together many times, and for me, it was like beginning life anew. We fell in love. I asked her to marry me, she said yes, and we married on December 16, 1978.

I had it all, everything a man wants and dreams of. She was lightning, thunder, sophistication, and upbringing. We shared the same values. There was early trust because we had known each other since she was in the first grade.

We were married in Knoxville, took our honeymoon in New York City, and split the time between the Plaza and Pierre Hotels. We shopped at Bergdorf and had many glorious lunches in New York's best restaurants. I don't think there were but two evenings that we didn't see a Broadway play.

She loved children. She loved old people. She gave generously. Carol had the strongest Christian conviction of any single person I ever knew. She was so much more than just a good influence on me.

I had always believed in God, but I had this notion that God might allow me into heaven one day simply because I was a good person. Or at least I felt that I was a good person—certainly not as bad as some I had met. I was proud of that and hoped that was enough. Carol helped me understand that there was more to it than I had thought. She was patient—and in the things that mattered, she was always right.

To me, she was everything. The beginning and the end. We would live our life together, grow old together, and be beside each other in rocking chairs one day.

The best day of our marriage was the day Dooley came into the world. I know every parent on the planet thinks their own children are the greatest thing since sliced bread. In a way I dreaded children because it would mean sharing Carol. I was in the delivery room when Dooley was born. The nurse laid him down next to me. We locked eyes, and in that moment, I looked at every feature of his face. He looked at me with intelligence and wonder and seemed to already know my voice. Even today I can still see every line in that bright new intelligent face. I cradled him in my arms and handed him to Carol, and we were a family.

We traveled together to Hawaii, Argentina, Costa Rica, Puerto Rico, Canada, Italy, France, Hungary, and Ireland. Some of our favorite vacations were simply low-key trips to the Florida Keys, to beaches in the Carolinas, or ski trips to the ski resorts of the West. In each place, Carol always found the very best restaurants and hotels, which I enjoyed immensely.

My days started and ended with her. I couldn't wait to get home from the office each evening, smell dinner cooking, and have Carol greet me with as much love as the day we were married in 1978. Then one day, after thirty-four years, it ended.

It was November 8, 2011, election day in Knoxville. We met at the Sequoyah School voting precinct and then went home for dinner. Carol was especially excited that evening because one of her longtime childhood friends, Julia Freer, had found an old film from Carol's first-grade birthday party at Carol's old family home on the farm between Knoxville and Farragut, Tennessee.

Julia had the old film converted to videocassette. After dinner, Dooley, Carol, and I went to the den and watched Carol's first-grade birthday party on the TV. The party had taken place mostly outside in the yard. We had stepped back in time in that video to 1950, the year I had met Carol, when she was in the first grade. And there she was, as pretty as ever, even though just a little girl.

Somehow, Dooley, Carol, and I felt a sense of sadness, seeing the old home place again, her mom and dad, sister and brother—and especially Carol as she faded away slowly in the last scene of the old movie.

Reluctantly, I pulled the video from the recorder and stored it carefully in one of the cabinets. Dooley returned to the office, and I opened my laptop in the kitchen to catch up on business emails.

Soon Carol passed by the door to the kitchen on the way to the den. She paused to say how much she appreciated the video from Julia and that she was going into the den to call and thank her.

Dooley and I emailed each other a few times about client projects. Finally, I closed my laptop and went to the den. Carol appeared to have fallen asleep sitting on the sofa. The phone was on her chest, and I thought she had dozed off without hanging it up. I walked over to the TV, turned it off and looked back to Carol. Suddenly a cold chill came over me, as something wasn't right. I couldn't see her breathing.

I rushed to her and felt her pulse. She was gone.

I grabbed the phone, called 911, and frantically spent the next fifteen minutes doing CPR while the ambulance came. After thirty-four years she had left me, and I never even got to say goodbye.

At first it was disbelief, as if there was some way to turn back time—and then the force of it hit me like the grief of a thousand years, and there was pain like I never imagined existed.

Looking back, I guess it had never occurred to me that she might die first. I was so sure I would be the first to go. I now realize that I was wrong about so many things.

Throughout our marriage, there was never a day I didn't tell Carol I loved her. But if you lose the love of your life in an unexpected instant, you yearn to have said so much more. I have often wondered what I would say to Carol if God would somehow grant me five more minutes with her—as if it would even be possible in five minutes to thank someone for a lifetime of love.

I would start by thanking Carol for believing in me. I would say, "You were my cheerleader and biggest fan. You bragged on me to your friends—you celebrated when I succeeded. You consoled me when I failed. You lifted me up when I was down. You sensed when I needed encouragement and always gave it—even when I was at my worst, you were always there for me."

I would say, "Thank you for the sacrifices you made—always putting family first. There were so many times I failed to notice, and yet amazingly, you never complained."

"Especially, I thank you for the way you guided our son. You lived for truth and set such a good example, leaving behind a beautiful legacy of love and kindness."

I would tell Carol, "You looked beyond my weaknesses, which were so many, and always pointed to the best in me." And I would say to her, "I still fall short, but I am so much better for having married you."

"You loved me when I was unlovable—when I had too high an opinion of myself and there was nothing that I gave in return—and yet you believed in me and encouraged."

"You did the thankless things. You washed my clothes, cleaned the house, cooked my dinners, shopped for me and more—the things I took for granted." I would tell Carol how much I now struggle with all those things, and I would say, "For the first time I realize how difficult all of this was for you. Yet amazingly you always made it seem so effortless—the dinners at home with close friends. The times when family gathered—Christmas, Thanksgiving and all the holidays—and always you strengthened bonds."

"You lived your life with more animation, love, and spirit than anyone I ever knew, and it rubbed off on all around you—and it was because of you that I learned how beautiful a life can be."

And I would say to Carol, "There was so much that we hoped for, things that might come true." And I would say, "When I think of all the good that happened—I owe all of that to you."

There is so much more that I would want to tell Carol, things I would need to say, and the magnitude of it overwhelms me. It would take another lifetime to say all the words that I will never get to say.

I wish that I could time travel.

PART FIVE

Transformation

DREAMING BIG

THERE IS SOMETHING about human nature that makes us want to improve and show progress regardless of our endeavors or goals. It's a lifelong need, and for the most part, progress comes slowly.

Take fly fishing for instance: I estimated that by the age of thirty-eight, I had made more than 1.5 million casts. I had started young and had fly fished often. I could do things with a fly rod without even thinking, like changing directions in the middle of a cast or throwing a fly into the teeth of a strong wind on a bonefish flat. Things that had been difficult in the early days of fly fishing became automatic, like muscle memory. It's something that begins to happen after a million casts, and I was living out my dream of catching a lot of fish.

By 1978, that dream had mostly been fulfilled, and I had begun to contemplate something more challenging—a dream of catching larger fish. For several years, the thought of it had simmered in my soul. A part of me was ready to raise the bar, to find out where the edge of the fly fishing world really was. I think it's a natural evolution in the journey of a fly fisher to want to do that—to advance from catching smaller fish to landing fish much larger.

I knew it might never happen if I continued fishing the same streams, the same way, year after year—especially the streams where more and more people were competing for fewer and fewer fish.

My goal would require a radical change—new places to fish. Frankly, I didn't yet have a clear sense of where those new places might be or where this dream of bigger fish might lead me—or how things might unfold.

SECRETS UNFOLD

FLY FISHERS ARE SECRETIVE about where they fish. It's a fact. The bigger the fish, the more they guard the secret.

Sometimes they will share techniques. Occasionally a good friend will offer general fishing tips or advice on the right flies or tackle. If you happen to be one of their closest friends, they might even tell you how they fish, but unless they are only hours away from death and have no heirs, they will never divulge where they fish. Never, ever.

I remember once when I was fly fishing in British Columbia with a guide who lived there. I had fly fished with him several times before. This guy was fun to fish with. He wasn't one of those 8 a.m. to 4 p.m. guides; if the steelhead were active, he'd stick with you until pitch-black dark, not because he had to, but because he loved it. One day he offered to show me "the best steelhead run in North America." He said, "I can have you there within an hour." Then he added, "Of course, you'd have to agree to be blindfolded for the drive to and from the river." I declined.

Much of this secretiveness was driven by necessity. Thousands of newcomers were entering the sport. The peacefulness and solitude of the rivers I had known as a kid were for the most part just a memory, and the beautiful streams I had loved and fished were slowly overrun by legions from the cities. I longed for the old days, the uncrowded rivers of my youth in the backcountry of the Smokies and in Montana, and I wondered if places like that still existed. Maybe those places were secrets, too?

That very secrecy is why I was so excited one Saturday morning in April 1978 when the new IGFA world record book arrived by mail. In the world record book, there were no secrets. Ironically, the record book revealed the place each fish was caught and the date.

I spent the entire day devouring every word in the record book, especially the information as to the location where these huge fish were hooked and landed. That's when the thought first occurred to me to use the world record book as a guide on where I might like to fish one day. I hadn't yet achieved the financial resources or success in business that it would take to journey to most of those places. But the thought of it motivated me to work hard so that one day I could do so.

As it turned out, some of those places were off the beaten path—rivers that were less crowded. Some were so remote, they offered complete and total solitude. Like the oceans of the world where sometimes you could go for days, or even weeks, without seeing another boat or angler. It was these places off the beaten path that appealed to me. It connected me with something I had lost along the way.

At this point in my fly fishing journey, my goal was not a world record. In 1978 I had not yet developed an interest in it, and frankly the idea of fishing for records did not occur to me until a decade later. I simply wanted to throw a fly at big fish. Really big fish. And then just see what would happen. It seemed to me that the places where the biggest fish were caught were right there in front of me on the pages of the world record book.

I wasn't much different than so many other fly fishers that had gone before me. Like them, I progressed from bluegill to trout, from trout to bonefish, and from bonefish to tarpon. All I had done was learn from them, or from their guides, or from the articles and books they had written. I felt like it was time to live and write my own stories—to push forward on my own. I longed to see how far that might take me and the places where it might lead.

Throughout the pursuit of this dream and the places it took me, I never lost my love of smaller rivers and intimate little trout streams. But most of these small streams and the rivers of my youth were now crowded with other fly fishers, and I yearned to explore places off the beaten path—the farther from roads or trails the better. On top of all that, I now had yet another reason to fly fish: my desire to catch larger fish.

To me, the rivers of the far north and the great oceans were the final frontier of fly fishing. And the fish were bigger, especially in the great oceans: marlin, tuna, wahoo, and sailfish. Maybe too big for a fly rod.

Or maybe not.

This was the 1970s, and the common wisdom of the day was that fish like marlin and sailfish were too powerful and fast for fly tackle. And beyond that, how in the world would you ever find one to cast to out in the middle of an ocean that large, anyway?

At the age of thirty-eight, I had fly fished for thirty-one years. I suspected that I had the experience and ability to begin the quest for big-game fish like marlin on a fly rod. I wouldn't have ever said such a thing to my fly fishing buddies at the time for fear of looking bad if I failed. But I clearly harbored the notion that if a really big fish could be caught on a fly rod, with a little trial and error, I could do it.

I wondered what it would be like to throw a fly at a marlin or some other giant game fish. I guessed it would be like moving up a weight class in boxing or in wrestling. You never know for sure if you are good enough to move up a level until you try. I was confident I had the skill set, but how would I ever know for sure if I could land one of those big fish unless I made the effort and took the first step?

Once that dream became a part of me, the thought of it kept pulling at me until I did something about it. That's why I brought a fly rod along whenever I was with family or friends fishing with conventional tackle on the ocean—just in case the opportunity presented itself to throw the fly at a billfish.

Back then, people thought I was insane, and the American charter boat captains discouraged me. In fact, some of them told me outright that I was nuts. One boat captain told me, "It's impossible. No one can land a billfish on a fly rod." He wouldn't even allow me to bring my fly rod on his boat.

These crusty old captains had their own way of fishing with heavy deep-sea conventional tackle. They got their bookings based on how many billfish they landed, and they didn't want some crazy angler with a fly rod on board losing fish that might hurt their standings. They wanted to come back into the dock each evening proudly flying a flag on their outrigger for each billfish they had caught that day. The clientele that came to the docks tended to charter the boats that flew the most flags.

I didn't let any of that bother me. The dream of fighting big fish on a fly rod was etched into my psyche, and I was powerless to fight it.

Gradually I began to leave the more crowded rivers and smaller fish behind, focusing more and more on the places where the biggest fish were caught. This led to amazing fly fishing adventures in some of the most scenic and secluded places on earth.

PART SIX

Memorable Casts

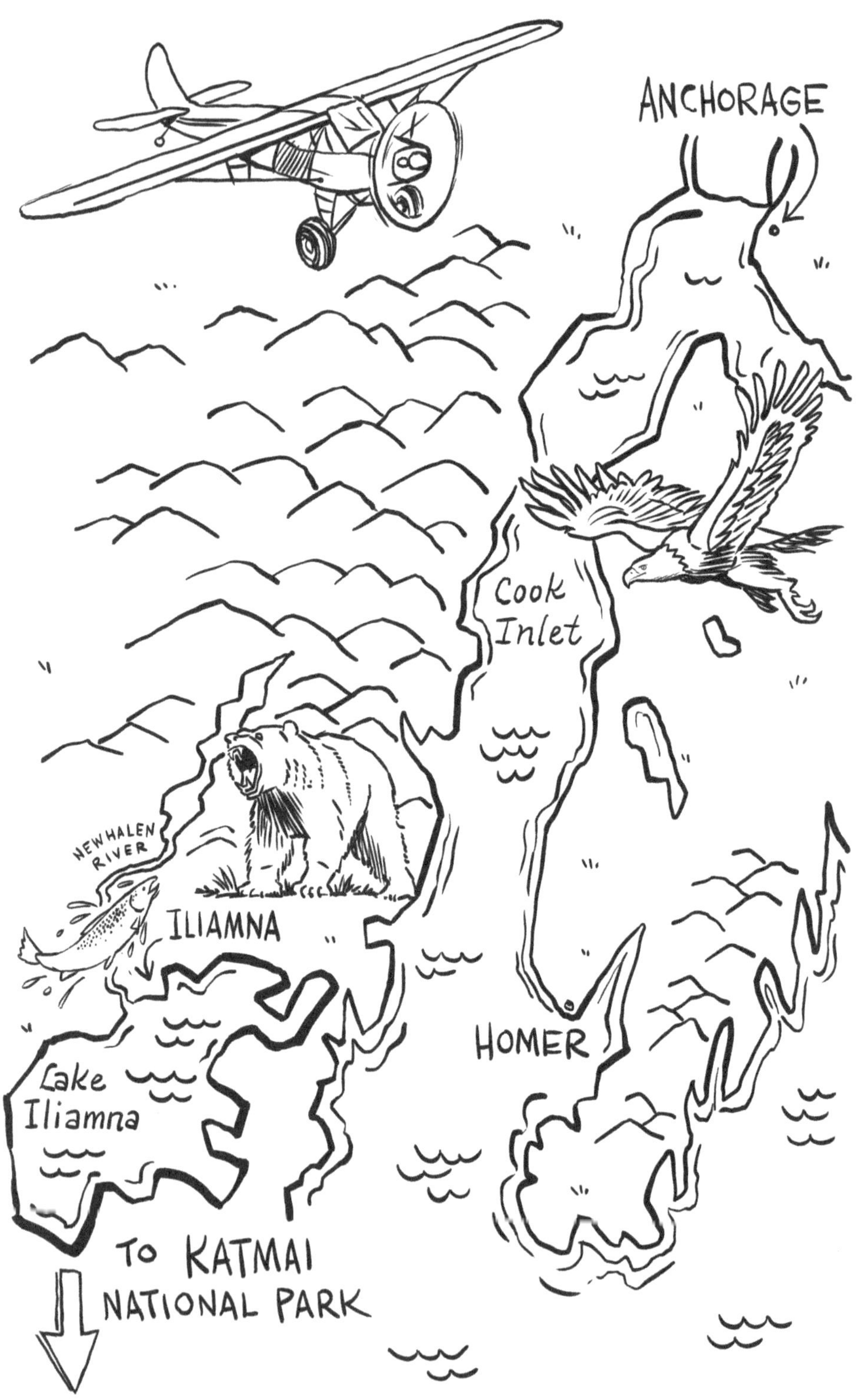
ANCHORAGE
Cook Inlet
NEWHALEN RIVER
ILIAMNA
HOMER
Lake Iliamna
TO KATMAI NATIONAL PARK

BROKEN LINE, 1988

THE LITTLE SUPER CUB was faded red and outfitted with oversized Tundra tires, so it needed no runway to land, only a few hundred feet of open tundra. I watched from the stream as it rumbled across the mossy terrain and lifted off effortlessly. The bush pilot was flying my fishing buddy, Kerry Sprouse, back to Iliamna and our base camp.

I had just turned forty-seven, and in all the years I had spent in Alaska, this had been one of the finest days ever on my favorite wilderness stream. Earlier this morning the bush pilot had flown the two of us, one at a time, to this majestic stream in his little Super Cub. It held only one passenger, so I had told him, "Watch the weather. If it starts looking bad, come pick up my friend first. I have a .475 revolver, tent, freeze-dried food, cookstove, and a sleeping bag, so you don't need to worry about me. If the weather turns bad and you can't make a second trip for me, don't worry—I'll be fine."

The fishing had been glorious. It was one of those days when all the stars were aligned. The river was filled with big rainbows. Kerry had caught a very large one, the largest I had seen on that stream, and it seemed that we could do no wrong. But now, I was alone, and the tundra was quiet. I remember the time. It was 4:45 p.m.

The wind had died, and there was nothing but the silence. Clouds and fog had moved in, almost in slow motion. I knew that it would be a miracle if the bush pilot was able to land the Super Cub here with near-zero visibility. And to me, it didn't matter; I was on one of the best rainbow trout streams in the world, and I had it all to myself. I waded out into an eddy. There in the long riffle to my front were some of the largest rainbows in Alaska. With no wind, conditions were perfect.

When I fish by myself, I have more time to experiment and try new techniques without the embarrassment of failure, and this was one of those days. I had always wanted to try extra-light leader with an egg pattern fly on a calm day. Extra-light leader sinks faster than heavier, stronger leader, and while it is a disadvantage to fight a large fish on light leader, I wanted to see if the faster sink rate of an ultra-light 6X leader would lead to more strikes.

I tied on a 15-foot leader tapered to 6X, attached the small fly,[8] and placed a tiny strike indicator on the leader about six feet above the fly. The strike indicator floats on the surface and helps you detect the strike.

Immediately I was into one fish after another. Most were in the six-to-seven-pound range. I was having a grand time, and it looked as if my crazy experiment with the 6X leader was working, at least on a calm day like this one.

It started snowing, but the temperature was holding steady, and there was no wind. In fact, the low cloud cover created a bit of darkness that seemed to activate the fish.

Occasionally I could hear the drone of an aircraft somewhere far above. Sometimes it descended closer and then retreated upward. I knew it was the bush pilot, trying to find a way through the snowy clouds and fog to land.

Silently I prayed he would not find a way in. The fishing was that magical, and I didn't want it to end. As soon as I landed a fish and hiked back up to the deep riffle, another would hit.

It was 5:30 p.m., and I heard the Super Cub no more. I made a long cast upstream near the far side and watched the tiny strike indicator ride the rippled current. When it arrived just upstream from me and about two-thirds of the way across the stream, the strike indicator stopped and slowly went under.

With a 6X tippet, you can't really set the hook. You just lift your rod tip gently, which I did. Something big was surging deep in the current. Then came a few seconds of violent head shakes. It was clear to me from each shake of the head that this was a monster fish.

---

8. The fly the author used was a small pink egg fly made from yarn. It represents a sockeye salmon egg, which the big rainbows in Alaska feed on.

He turned and headed downstream toward the mouth of the river at a high rate of speed. He never jumped, but when he neared the lake, his head appeared above the surface and he continued his run, lunging on his side like a marlin with his head out of the water. He was well over a hundred yards into my backing, and my rod was pointed straight down the line with the reel on total freespool to minimize drag. No matter how fast I ran downstream in the current, he was going faster, and the line on my reel was getting thinner.

I clambered out onto the bank and ran through the tundra to the mouth of the small lake still losing line to the fish, and he was now about a third of the way across the lake.

I waded quickly out into the water as far as possible to shorten the distance between me and the fish without really daring to put any real pressure on him this early in the fight.

Instead of continuing across the lake, he took a right turn and ran parallel to the shore on my side of the lake. The major problem here was the large and long bow in the line dragging though the water as he changed direction.

Eventually I was able to catch up with him and shorten the line as he moved farther down the lake. Soon he got tired of that and decided to run all the way back up to the mouth of the river. I thought maybe that was in my favor, because now he was fighting the current and me.

After about five minutes, I decided to move closer and try to gain some line. I never came close to getting any of the hundred-foot fly line back on the reel when he must have sensed me coming and took off again, running even farther upstream and way into the backing on my reel.

I clambered up the bank and ran upstream to the riffle where I had hooked him. Suddenly he turned and took off on the fastest run so far, heading downstream again on a scorching run and not even slowing until he passed the mouth of the river and was halfway across the lake.

By now I was tired of running but too excited not to. Once again, I climbed the bank and ran through the tundra to the mouth of the lake. With my rod still pointed at the fish, and with no drag on the reel, he was making another long run past the middle of the lake. I let him run pretty much at will until, at last, he slowed and stopped.

At this point I decided it would be best to fight him in the lake, where there was no current. I wouldn't be able to get as close to him in the lake as

I could if he were up in the stream, but in the lake, with no current, I felt I had a better chance of landing him.

The shore was shallow with gravel point bars running out into the lake just under the surface. They looked to me like wonderful places to land a fish. I knew the biggest danger in the lake would be if he ran all the way to the outlet. If he made it to the outlet at the far end of the lake, the water would be deep and the current too strong to land a fish of this size on such a light tippet.

For the next hour, the big rainbow made repeated runs out into the lake and down a bit toward the outlet. At the end of each run, I put light pressure on him and tried to lead him back toward the shallows and farther back up the lake, away from the outlet. Every time he felt the shallow gravel, he would turn and make another run toward the middle of the lake. However, I noticed that after each of his runs, I was able to work him into water that was shallower than the time before.

I was gaining confidence that I could land the fish if I could just do it before we arrived at the outlet of the lake.

His runs were getting shorter now. At no point during the fight did he ever leap clear of the water. I think maybe he was too large to want to do that, although he surely could have.

Pretty soon I had him on a gravel point bar in about a foot of water. A good portion of his back was now out of the water, and he was huge. But each time I worked him into shallower gravel and his tail touched the bottom, it would spook him into another run out into the lake.

I didn't have a landing net; I never use one in Alaska if I am wading. But it didn't matter—he was too big for a landing net anyway. My plan was to get him up shallow enough on a gravel bar to where he would eventually turn over on his side and lose the power of his tail.

We were getting close to the outlet of the lake, but his runs were shorter, mostly less than fifty feet.

There was only one more long gravel bar between me and the outlet.

I didn't want to go that far unless I had to. While I was eyeing that last gravel bar, the fish turned toward me. I began backing up with the fish swimming right into the gravel in front of me. At the last second, he felt the gravel on his belly and panicked, but instead of turning and making another run to deeper water as before, he ran sideways toward even shallower gravel. Now he was upright, and about two thirds of his body was completely out of the water. I tried not to think about how enormous he was. I needed total calm and concentration. He could still escape if he could turn.

To discourage that, I maneuvered around to the deeper water so that I was between him and his escape route. He saw me and started trying to swim away, but each stroke of his tail was taking him farther up the gravel toward the shore. Finally, when he was in about four or five inches of water, he turned over on his side.

I hadn't realized how mammoth this rainbow was until then. I slipped the tape over him. He measured thirty-four inches in length with a nineteen-and-a-half-inch girth.

I checked the time. It was 7:15 p.m. I had fought this fish an hour and forty-five minutes. I knew he was somewhere in the fifteen-to-twenty-pound range. The thought occurred to me then that he might be a world record rainbow on the light 6X tippet, but I was determined to release the fish—except first I wanted a photo.

The water temperature was so cool on this late October evening that I knew he would survive a quick photo in the shallows. I fumbled around in my vest and realized I had forgotten to bring my camera (this was in the pre-mobile phone days).

I thought for a minute and remembered the pilot had a camera in the Super Cub. That wouldn't be of much help until the next day. What I needed was something like a stringer to hold the fish alive overnight, and when the pilot came back in the morning, I would get the photo and then release the fish. But of course, a stringer is the last thing I would have with me here in the backcountry or on any other day fly fishing.

I cut off about fifteen feet of fly line from my reel and ran one end down through the monstrous jaws, along the inside of the gill plate, out the bottom, and back around the outside to three or four inches beyond the front of its huge kype, where I tied it back to the fly line. I was careful that nothing pinched, pressed, or squeezed against the mouth or touched his inside gills. I then towed the fish, slowly at first, until he regained strength in this cold water, checking on him frequently. He was recovering nicely.

My plan was to tow him back up the lake and then up the river to the place where I had first hooked him, and once there, I would tie the fly line to a stout willow bush hanging over the river. It was a good place for me to pitch my tent for the night.

I worked my way in waist-deep water back up to the mouth of the river. I heard him snapping his jaws behind me only inches from my waders. His jaws were so large, I nearly feared he might bite a hole in my waders, so I gave him a bit more fly line to keep him farther away as I waded.

Arriving upstream at the riffle where I had first hooked him, I was amazed at how much strength the fish had regained. Being towed had put a lot of water through his gills at just the right speed. I was about to tie the fly line to the base of one of the largest willow bushes leaning out over the river, but two thoughts changed my mind.

First, if a grizzly were to come along during night and the fish was on a short line, it would be too easy for the bear to catch him. Second, I was concerned that the fish was so powerful, he might be able to break a single strand of fly line.

So I cut off another sixty feet of fly line from my reel and doubled it into one thirty-foot length. The core of that Scientific Anglers fly line was 20-pound test. By doubling the fly line, I made it a 40-pound test line.

I ran that doubled line through the jaws of this beast and tied it as before. Then I took the other end of the doubled fly line and tied it to the trunk of the willow. The fish was now tied to thirty feet of 40-pound test.

I knew it would be dark soon, and it was still snowing. I pitched my tent a hundred feet back from the stream on some level tundra and put the sleeping bag inside. Still in my waders and fleece for warmth, I found a comfortable spot outside the tent and fired up the little cookstove.

I cooked hot chocolate and a freeze-dried meal pack. It was turkey tetrazzini, and it was delicious. Topping it off with a cup of coffee, I leaned back against one of the tussocks in the tundra and watched it snow. It was surreal—the mystique of it seemed to cast a spell. And I felt like maybe God had made this moment and this view just for me. And it began to sink in that I had just landed the rainbow I had fished for half my life, and now, I was hundreds of miles from civilization. I was alone in the snow, and I was not cold. I thought of my fishing buddy back in Iliamna. I knew he was probably concerned about me but confident I would be fine.

When I turned in, there was three or four inches of snow on the tundra, and still there was no wind. I placed my wet waders and wading shoes inside the tent to keep them from freezing overnight.

The next morning, when I poked my head from the tent, the sun had just appeared. It was cold, well below freezing, and still there was no wind—unusual for this place. I took a deep breath of the cold Alaskan air and watched it condense in a huge cloud as I exhaled.

Soon I was in my waders, mostly for warmth, and I laced up the felt-sole wading shoes. I added a heavy fleece jacket. My first thought was of breakfast, and then I remembered the fish, and I couldn't wait to have a look at him in the sunlight. I hoped that a bear hadn't got him. He was the largest rainbow I had ever caught, and I wanted a photo of the fish before I released him.

As I walked across the tundra toward the river, my heart surged with joy. I could see the willow tree wagging back and forth, pulled strongly by the doubled fly line and the fish unseen below in the current. It meant he had completely recovered overnight and had not been eaten by a grizzly.

Eager to see this fish of a lifetime, I walked straight toward the willow. When I was within fifteen feet of it, which was about forty-five feet from the fish, he spotted me in the bright sunlight. It spooked him, and he tore across the river like an F-15, the line behind him making an accelerating and sizzling sound as it ripped through the current. Reaching the end of the doubled fly line, there was a loud crack like a gunshot as the fish broke it and raced across the river.

Never in my wildest imagination did I ever think a rainbow trout could break a doubled fly line. But he did, and he was gone. I stood there alone on the tundra in shock.

After a few moments of disappointment and disbelief, I walked over to the willow, untied the remaining fly line, and stuffed it in my backpack.

The exact weight of the fish I will never know. But from the measurements I had taken of a thirty-four-inch length and a nineteen-and-a-half-inch girth, I estimated that the fish was eighteen pounds.

It is still to this day the largest rainbow I ever saw or landed, and I am often amazed that I did it on 6X tippet in a snowstorm, and so far from roads or trails.

Later in my life, I caught even larger brown trout in Argentina—several of them, actually. But this rainbow, at one hour and forty-five minutes on 6X, was in my mind the pinnacle of all my years of trout fishing and the greatest fight in my life with a trout on a fly rod.

N
W
E
S
BRITISH VIRGIN ISLANDS
HAITI
DOMINICAN REPUBLIC
Caribbean Sea
ARUBA
THE MARINA
LA GUAIRA
CARACAS
VENEZUELA
COLOMBIA

MAN OVERBOARD, 1994

MOST WRITERS AGREE that there are four stages in an angler's life. I happen to believe there are five or maybe even six. But let's assume, for the sake of argument, that there are four, the first being that you want to catch a fish.

My first fish on a fly rod was a bluegill in 1947 at the age of six. I then graduated to stage two, which is where you want to catch a lot of fish.

The third stage of a fly fisher's life is when you want to catch a big fish. In 1994 I was well into that third stage. In fact, my goal was not to catch just any big fish.

I wanted to catch the toughest billfish in the ocean, a blue marlin, and I wanted to do it the hard way, on a fly rod. It never occurred to me that if I succeeded in landing a blue marlin on a fly rod, I might then want to catch a lot of them—which is stage four: wanting to catch a lot of big fish.

A blue marlin is considered the fastest, strongest, and meanest game fish on the planet, and at the age of fifty-two, I had been nursing the dream for several years of landing one. I wondered what would happen if I got the opportunity here—so far from home, where I had never been? Could I make it count? By now, I had already landed well over three hundred sailfish on a fly rod and felt ready for something more.

At the time of this story, only two people in history had landed an Atlantic blue marlin on a fly rod. The first was Billy Pate. His fish, a world record, was ninety-four pounds. It was later surpassed by a 113-pounder caught by

Harry (Jim) Gray off San Salvador in the Bahamas. Both men were great fly fishermen. Both were good friends of mine, and both are now deceased.

The drive to catch a blue marlin on a fly rod had taken me south, way south, to a place that held great promise. It was late May in Venezuela, and the morning view of the vast Atlantic was breathtaking from the open-air restaurant a few hundred yards from the beach. As I paid for my breakfast, I nearly handed the senorita my good luck coin, but caught myself just in time. It was a Bahamian coin, one my dad had brought back from Bimini and had given to me, and it had a blue marlin on it.

Even the parrot that was perched by the cash register bid me good luck in Spanish.

"Buena suerte," said the bird.

"Gracias," I replied.

Each morning, I looked forward to hearing that bird talk, but to a guy from Tennessee, it was always strange to hear it in Spanish.

I made my way down to the marina, where the sport fishing boat I had chartered was docked. It was a good boat for fighting a big fish—not too long, only forty feet or so, but the rear deck was spacious with plenty of room to move around and fight standing up with a fly rod.

La Guaira Bank, off Caracas, Venezuela, was one of the best places in the world for billfish, especially in February through May. The seas were often rough, but the wind tended to die down a bit in the afternoons.

We fired up the engines and left the dock early for the long ride north in high seas, setting a course for the fishing grounds. I passed the time by snelling hooks for my tube flies and tying spare leader tippets with Bimini twist knots. It was no easy task in this powerful sea with the boat swaying and lurching violently.

The sun was still low in the east as we eased down to trolling speed, lowered the outriggers, and let out three hookless teaser lures that were soon plunging and darting enticingly through the high waves of blue. We fought a strong northeast wind all morning under a blue sky and raised no fish. At noon, the stout morning wind began to weaken slightly.

Finally, at 1:30 p.m., a fish rose to the left long teaser lure. I couldn't see what it was, but the mates, Oscar and Wilfred, were yelling, "Azul! Azul!" at the top of their voices.

I grabbed my fly rod, and then, with the sun behind me, I saw him, and it was almost frightening. He was huge, and every stripe on his massive body was bright. He was majestic blue, and he looked angry. This radiant beast was coming in strong and fast, crashing the hookless teaser lure repeatedly.

Wilfred quickly reeled in the other two teasers to get them out of the way, and Oscar teased the big blue to within twenty-five feet of the boat. I could see him all lit up and couldn't believe how big he was. Every stripe on his body glowed like neon electric against the cobalt-blue depths below. He looked muscular, and his movements were quick and purposeful, like he ruled the ocean.

The captain, Julio, had been steadily slowing the boat, and then he put it in neutral. My heart was pounding as I cast the fly. It landed in front of the marlin. He struck it with his sword as he overshot the fly and failed to get it in his mouth. He then wheeled around and viciously took the fly going away. It was the most explosive and ferocious strike I had ever witnessed in all my years of fly fishing. I had seen violence in my life, but nothing like the viciousness with which this marlin attacked the fly. There was no need for a hook set. It was just fish on, hang on, and don't mess up.

The big marlin accelerated quickly across the surface of the Atlantic for nearly five hundred yards, going straight away with a slight right angle. He jumped again, again, and again, and when he burst through one of the tallest waves and into the air, and we were low in the trough, he appeared almost higher than the boat. The handle on my reel was spinning in a blur, like a turbo prop. To have even touched it accidentally would have resulted in a broken knuckle or finger.

As the run progressed, the marlin's great speed decreased a bit, and the jumps became more like greyhounding. They were long, high-arching jumps, and on some of them it was almost as if he was hanging in the air—which worked to my advantage, as the marlin's dorsal fin was less apt to cut my leader on those types of long arching leaps. I had backed off on the drag all the way to freespool, always using the palm of my hand ever so slightly on the bottom of the reel to prevent line overrun.

Julio was backing the boat aggressively, and I slowly gained line to where the fish was only about a hundred yards behind the boat. Then the marlin made two more short fast runs of about four hundred yards, each punctuated by the same high, greyhounding leaps he had made earlier. After that, I was gaining line when I realized he had suddenly gone deep. We backed the boat down on him as the line became tight. I estimated he was forty yards into the depths below.

I engaged some drag and put about eight to ten pounds of steady pressure on him. There was nothing more I could do except maintain constant pressure and angle. Now that he was deep, he spread those big pectoral fins (side fins) to catch the current and to make me do all the work. I leaned and strained against him and the strong, deep current of the Atlantic. I felt like I was trying lift a Boeing 727. I dared not force it, or I would break the tippet. Nevertheless, I had to exert a lot of muscle to keep a constant pressure on him.

I needed to pace myself and not wear out before he did. My friends Billy Pate and Harry Gray had warned me that it would be like a game of chess. They had said that sometimes the marlin makes a power move and sometimes it's a speed move, and sometimes the move is deep to see who wears out first. They had cautioned me that without constant pressure, the marlin would just swim deep, and rest, and never quit.

I applied even more pressure, but I never gained more than ten or fifteen feet of line, and each time he would take it all back. He stayed deep, swimming steady for a solid hour. My muscles ached and I felt the wear of the fight. Suddenly the line came to life as he changed strategy and made his next move. I backed off on the drag, and just in time. He came up fast, leaping clear of the water and crashing the surface about ninety yards away.

Then he tore off on a violent run, jumping repeatedly and erratically, slamming and smashing the surface of the ocean in ways I had never seen. At last, he slowed somewhat, which allowed Julio to back the boat down and put me in a position to fight the fish more closely.

I put a lot more pressure on him. After another hour, my legs and back were tiring, but we were even closer to him, and it gave me hope. Like a prize fighter who had never been into the tenth round, I had never fought a fish this big for so long, and I leaned back heavily against the power of the fish. After another thirty minutes, I was able to work the marlin just under the surface on the port side of the boat. Wilfred and Oscar were ready with their eight-foot gaffs. Wilfred leaned out over the gunnel and swiped aggressively at the marlin with his gaff, hitting the fish back toward the tail. In the blink of an eye, the big marlin violently pulled Wilfred across the wet deck of the boat where he crashed loudly into the transom—and the gaff pulled out. I wondered if Wilfred's ribs were broken, but he was tough and never winced.

After another grueling hour, with legs that were wobbly and arms that were bursting, I worked the fish back into position on the port side. Wilfred took another gaff shot at the fish. The ocean exploded as the marlin felt the gaff, and the next thing I knew, the fish was pulling Wilfred through the air, way up high and over the side of the boat. In the process, the gaff pulled out a second time, and Wilfred, with eyes as wide as saucers, was swimming in the Atlantic.

The marlin was still hooked up. I loosened the drag, as I didn't want to risk breaking the line during the melee. I figured the first thing to do was help get Wilfred back in the boat and not lose the fish while doing so. Julio was backing the boat down to get close to Wilfred, who had been pulled by the marlin a considerable distance from the boat.

Fortunately, the wind had lain just a bit. However, the waves were still high, and we had to be careful that the boat would not come crashing down on Wilfred.

Julio backed around cautiously to where we could ease in close to Wilfred on the starboard side. Oscar reached down with one hand, and I held my rod in my right hand and reached down with my left. Somehow we each connected with Wilfred. As he rode the crest of an incoming wave, we hoisted him over the gunnel and onto the back deck of the boat.

There was no time for high-fives. Line was flying off my reel as the fish took off on another fast run. We followed as best we could, always losing more line, and then at last the fish slowed and went deep again.

He was deeper this time, and I wondered how long this fight would last, and more importantly, how long I would last. The next hour and a half was brutal, and he was in total control. I worked hard to gain a little line, and he showed his power by immediately taking it all back and then some.

Suddenly, I felt him come to life. I started reeling fast and loosening the drag on the reel, because he was coming up quick, making a chess move. He bolted through the surface of the Atlantic and crashed down surprisingly close. I was at the perfect angle as Julio set the boat to run parallel to him in reverse.

The marlin was swimming fast and off the port side. For the next half hour, I worked him closer and closer, with me standing well back on the starboard side. At last the fish was almost there, but not quite in gaffing range. I applied all the pressure the tippet would take. At first, he wouldn't budge, but after another ten minutes, I slowly leveraged him closer and closer. Then came the important moment. Both mates took gaff shots at the fish simultaneously. The marlin exploded on the surface, sending a giant wall of water into the air and the boat. At first I thought both mates were going overboard. But they barely hung on, and the gaffs did not pull out. The struggle was violent as Julio scurried down from the bridge. It took the three of them with all their strength to hold on and then to pull the big marlin over the gunnel of the boat and onto the deck.

It was massive. I was in a state of shock. My legs were trembling from the long fight. I had never fought a fish for that long standing up, and my legs were like rubber as my heart pumped with excitement. I was in a state of disbelief.

I had dreamed about landing a blue marlin on a fly rod for so long that I could scarcely realize the magnitude of what had just occurred. While I didn't know exactly how big it was, the fish was clearly well over two hundred pounds. Suddenly the enormity of it hit me.

My goal had been to simply land a blue marlin on a fly rod, but I had also just shattered the 113-pound blue marlin fly rod world record, which my friend Harry Gray had set in San Salvador. As I stared in awe at this beast, the realization set in: the first billfish in the Atlantic over two hundred pounds on a fly rod. It would also put me in the IGFA's prestigious ten-to-one club, which means the fish weighed more than ten times the breaking strength of the line.

The four of us were stunned. We sat there together on the deck of the boat with the huge fish and celebrated. The radio was going nuts with other boats and anglers calling in congratulations. Word was traveling fast. Messages were relayed to us from as far away as the Florida Keys.

I am not a superstitious person, but as I sat there on the wet deck of the boat by the head of the big marlin, I reached into my pocket. And there it was—the blue marlin coin. I thought of my dad who had brought that coin back with him from Bimini, how he had taught me to fly fish, and all the times we had fished together before he had passed away. I wished that he were here with me now to celebrate. And from that moment on, the coin was always in my pocket, whenever I fished for billfish.

After a victory cigar, Julio fired up the twin diesels, and we were headed back to the dock in La Guaira on seas that were by now considerably calmer than when we had embarked that morning. It was almost as if the world had somehow changed.

On the long hour-and-a-half run to the dock, I was busy studying the IGFA world record application form and meticulously recording my equipment on it, which included the 20-pound test Mason tippet, Fisher 14-weight fly rod, Abel 5 reel, Ande 100-pound test shock tippet, Cortland fly line, and my custom-tied tube fly with two hooks, the front hook being an Owner 6/0 and the trailing hook a Tiemco 4/0.

When we arrived at the dock, a very large crowd of Venezuelans had gathered for the weigh-in. It took thirty minutes to get the big fish lifted from the boat up to the dock and onto the scales. I was stunned when the fish weighed in at 221 pounds.

The scales were accurate. However, they were not IGFA certified, which is required for a world record. A call went out to all the sport fishing boats in Venezuela to see if anyone had scales that were IGFA certified. As luck would have it, there was an American sport fishing boat on the way from the Florida Keys with IGFA-certified scales. Onboard was Enrico Capozzi, the owner and renowned angler. He expected to arrive the next day and said he would be happy to loan us his scales.

I decided to wait the extra day and weigh the fish on Enrico's IGFA-certified scales, even though it would certainly result in some weight loss due to the fish dehydrating overnight. So we took the fish to a refrigerated distribution company with a large walk-in cooler.

The next day Enrico's boat arrived with the certified scales. We hung the fish once more at the dock and reweighed it on Enrico's scales at 208 pounds. The Marlin had lost thirteen pounds overnight as it dried some in the cooler. However, since Enrico's scales were certified, I elected to enter the fish to the IGFA at the reduced weight of 208 pounds even though it was actually a 221-pound fish the day it was caught. Enrico helped me complete the world record application form, which was later approved by the IGFA.

Amazingly, at the time of writing this in early 2022, this world record, set on May 20, 1994, has stood for nearly thirty years. The record was rated sixty-third in the top one hundred world records of all time in the June 2013 issue of *Sport Fishing* magazine. It was the first time in history that an Atlantic billfish over two hundred pounds had been landed on a fly rod. Suddenly I was thrust into the fly fishing scene with coverage in the major outdoor sporting and fly fishing journals. I received letters, notes, autographed books, and messages of congratulations from people I admired, like Lefty Kreh, Joan Wulff, Guy Harvey, Flip Pallot, Billy Pate, Harry Gray, and more. It was the first of eight fly rod world records I would set between 1994 and 2002.

I had dreamed of landing big fish on a fly rod, and now I was finally fishing the oceans of the world and pushing the boundaries of fly fishing.

I had achieved stage three in the life of an angler, and I thought there was nothing more.

Lake Iliamna
COPPER RIVER
KOKHANOK
GIBRALTAR LAKE
DREAM CREEK
UNKNOWN CREEK
Cook Inlet

NIGHTMARE AT DREAM CREEK, 1995

COUNTING THE PILOT, there were three of us in the little Cessna. We were flying high above the largest lake in Alaska.

I was listening to the deep and heavily throated roar of the single engine when it quit. It didn't sputter and quit. It just quit abruptly. The silence was startling.

I am not a pilot, but it seemed to me that a thousand feet in the air and seventeen miles from shore is not the best place to lose the plane's only engine. I felt a tightening in my stomach.

I glanced at my buddy John, who was already keenly aware. The bush plane pilot, Tim, seemed unrattled and was methodically checking gauges, rapidly checking one instrument and then the next. To me, each second seemed like eternity.

I watched him go through his mental checklist. He flipped a small silver switch on his dashboard, and the engine on the Cessna 182 roared back to life.

Tim simply turned, smiled, and said, "We lost a magneto. Fortunately, we have a spare, which we are now running on."

I breathed a sigh of relief and gave a look of reassurance to John that was far more "reassured in appearance" than what I felt.

We were headed south across Lake Iliamna to a place I had always heard of and had wanted to fish. It was called Dream Creek. Perhaps it was a name too good to be true, and maybe I should have taken this incident as an omen that sometimes the things we dream of end up being different than what we expect.

Our flight path to Dream Creek from the Native village of Iliamna was south, over the big lake, and then across the famed Copper River.[9]

As we flew high above the Copper, Tim banked the plane to the left. We went in lower with a glorious view of this beautiful river, where I had fished so many years with my Knoxville friend Sam Patterson.

We flew the length of the river, from the big lake up to where the rapids began. Then Tim banked right, cleared some high ground, and we swept across the wild tundra.

At last, we were above Lake Gibraltar, and finally the stream itself came into view. Unfortunately, another Cessna floatplane was already parked at the mouth of Dream Creek. Whoever it was, and wherever they had flown from, they had beat us to our destination.

Dream Creek, although a nice size, was not a large enough stream for two parties of fly fishers at one time, so I asked Tim to look for another stream nearby. Only a few miles down the beach, a much smaller and unnamed little stream flowed into Lake Gibraltar. I asked Tim to land there.

Tim said, "I've never seen anyone fish that stream. It looks like a good one, but you know, sometimes things here in the Alaskan bush can surprise you." I thought to myself that was an odd thing to say, and I wondered why no one fished it. Was there a reason?

I asked, "Does it have a name?"

Tim replied, "No. Not that I know of. Maybe you can name it."

It was about half the size of Dream Creek. The only downside that I could see, other than the smaller size (which probably meant smaller fish), was that it ran mainly through heavy willow trees that would make it tough to cast.

In the unlikely event that Tim couldn't make it back to pick us up later that afternoon, we had brought two tents, a stove, fuel, bedrolls, and food.

However, we would need an open spot to camp.

The beach near the mouth of the creek was wide and open—a perfect spot to camp—but I prefer not to camp on beaches in Alaska, since bears

---

9. Several Copper Rivers exist in Alaska. The one referred to is east of the Native village of Kokhanok and south of Lake Iliamna.

use them as highways at night. I asked Tim to see if he could circle around and help us scout some high dry ground out of the willow trees and away from the beach, in case we had to set up a camp. We soon spotted a level promontory on the side of the small mountain nearby.

Tim circled and landed the plane on the still water of the lake and taxied into the dark, brown pebble beach. We quickly unloaded our gear, and moments later, Tim and his plane vanished into the blue horizon.

I love exploring new places, as did John. We were excited to get started. We stashed our camping gear high on the beach at the base of a large sand dune and walked anxiously toward the mouth of the creek.

As we approached the stream, I could see no sign that any human had ever been here. It looked so inviting, and yet there was something about it that didn't feel quite right. That feeling evaporated when I spotted movement of twenty or thirty sockeye salmon on their spawning beds. This most assuredly meant there would be rainbows here feeding on the sockeye salmon eggs.

We rigged our 6-weight fly rods with 12-foot leaders, strike indicators, and egg patterns.

John Emert is a great fisherman. Straight out of the Smokies like myself, he had grown up in Townsend, Tennessee, on the edge of the national park, and there is no one I would rather share an Alaska trout stream with. We had fished together so many years on the remote streams of Alaska that we could nearly communicate without speaking. That morning, because of the diminutive size of the stream, we alternated pools.

John was first up and immediately caught a very nice rainbow. As it turned out, there were fish that size in every riffle, and we fished on and on, losing all track of time, and before I knew it, it was already afternoon.

We moved silently and came to an area of the stream where the large dense willows blocked the light. Here the stream was narrow, not more than fifteen feet wide. The banks had closed in on each side, and the willows hung over the stream like a tunnel, blocking the sky, and inside that tunnel it was almost dark. The vegetation on each side of the stream was so thick, visibility was only a few feet.

To me, this narrow and dark place seemed foreboding, but the fishing was good, and with John only a step or two behind me, we waded into this eerie place. Before me in the murky dimness, I saw a deeper pocket of water that might hold a fish. I was using an old Orvis CFO fly reel. I loved that reel because when I hooked a fish and the fish made a run, the reel literally screamed in a high-pitched whine. It was an electrifying sound and one I liked to hear often.

I stripped some line off my reel to make a cast. It caused the reel to make that high-pitched whine. That's when the world exploded, close and all around, like an earthquake or a herd of elephants stampeding in panic. It happened all at once, giant pounding of feet, the ground shaking. Trees were being ripped from the earth, crashing down in every direction.

Then it was quiet, except for my heart, which I could hear pounding in my chest like a hammer. Slowly the realization and the chill set in.

I looked at John, and in this moment of epiphany, we knew we had just walked into one of the largest dens of grizzly bears in Alaska.

My first thought was to get out of there—and fast.

Still absorbing the shock, we weighed our two options: go upstream or go downstream. We reckoned that we were closer to the mountain upstream than to the lake downstream. We chose to keep moving upstream.

I felt the tightening in my stomach for the second time that day.

Soon we came to a place on the left where the thick vegetation opened up a bit. We caught a glimpse of the more open mountainside beyond. Up there was the promontory we had scouted from the airplane. We made a beeline for the mountain, and soon we were making the steep climb.

Once on the promontory it was like a huge weight lifted from our shoulders. I knew it might be short-lived, and we couldn't stay long. We needed to be on the beach when Tim returned in the Cessna. Even if he didn't make it back to pick us up, we would need our camping gear, which was stashed there.

At least for now, here on the high ground, there was visibility, and nothing could attack unseen. Below us, we could see the tall underbrush and the entire valley and stream feeding out to the lake.

But then a cold revelation set in, and what unfolded below was, once again, chilling. I said, "John, there's a grizzly right where we left the stream."

He replied, "There's a sow and three cubs down by the lake."

"Look! There's another just below us," I reported.

"And two more by the second bend in the river. And look out there a hundred yards on the left of the stream. There are three more," said John in amazement.

And so it went. In less than one minute we counted twenty-eight grizzlies, and somehow, we were going to have to walk through all those bears to make it to the place on the beach where we had stored our gear.

In the Alaska bush, there is a rule. It is called *the rule of twenty-seven*. Most of the guides, natives, and people who live in the outback know of it. Someone even wrote a book about it, which I read a long time ago, but I can no longer find it. Basically, it's a prediction of when you will be attacked by a grizzly.

The book was based on actual Alaska grizzly bear encounter data. The premise of *the rule of twenty-seven* was that on average, every twenty-seventh grizzly bear you come upon will attack you. You are lulled into a false sense of security when the first twenty-six grizzlies you encounter don't. Then, you meet bear number twenty-seven.

The rule weighed heavily on my mind. John and I had spotted twenty-eight grizzlies, and that was just the ones we could see. How many more were hidden in that tall grass?

We needed protection. That's why John and I each carried .475 revolvers. We had tested them with .425-grain bullets at 1,275 feet per second. We carried our revolvers in holsters strapped around our waists much like they did in the old west. But now, as we ventured into the thick willows and tall grass, we carried the guns in our hands at the ready.

We were headed for the place on the beach where we had stashed our tents and gear. It was tough to see in this tangle of tall grass and thick willow trees. How many more grizzlies were concealed that we had not seen from high on the promontory?

I have heard it said that in a wilderness, you are either the hunter or the hunted, the predator or the prey. I wondered if the .475 revolver I held in my right hand put me somewhere in between. Here in this tangled morass of tall grass and willows, we could see only a few feet. An attack by a grizzly would come quick, with only a second of warning, two at the most. No time to think. Just react, point, and maybe get one shot off.

These were Alaskan grizzlies that averaged eight to twelve hundred pounds. The chance of a shot killing one of these brown beasts in less than a second or two during a charge at close quarters was not zero, but it was about as close to it as you could get.

Still, the .475 felt reassuring—not like that day in Montana when I had left the gun in the truck. At least this time I had it with me, and John had one, too. So maybe we get two shots off. But it was problematic at best. I felt the tightening in my stomach for the third time that day. I thought, "It's Vietnam again. Different enemy, but one that could be waiting in close ambush."

Thoughts like these filled my head as we bushwhacked through the labyrinth. This Dream Creek outing was not going the way I had planned. I had envisioned an idyllic braided stream meandering across open Arctic tundra. We would be there alone with twenty miles of shoals and runs filled with rainbows. But my dream was not reality. Instead, we were here, on a no-name creek, in Alaska's version of a southeast Asian jungle, and we were not alone.

And how many bears were we passing in the thick brush that had not attacked? Would the count reach twenty-seven?

Finally we saw a small opening ahead. We broke out of the thick vegetation and crawled up the back side of one of the tallest sand dunes and peered down the beach. Here came a huge bear, and he was headed straight for us. Or was he headed straight for our camping gear stashed just below us?

The big grizzly sniffed around, took the blue bag containing my tent in his mouth, and tore it open. He stood up on his back legs and shook it mightily in his mouth. Then he sprinted down the beach with my tent in his mouth. It was shocking how fast he could run. He ran about a quarter mile, then turned and ran back. He ran past us, my tent billowing like a streamer in the breeze.

After several runs up and down the beach, he decided to test his tent-swimming skills. Pulling my tent behind him, he swam two hundred yards, straight out into Lake Gibraltar. Then he swam parallel to the shore back and forth. Finally he swam to the beach right in front of us, where he decided to end that part of his performance by furiously shredding my best backpacking tent into pieces.

Then he came toward us again, to where the rest of our gear was stashed. He grabbed the can of Coleman white gas and crushed the top of it with his jaws, even standing up on two legs and lifting it up to drink. I would have given anything for a match and a way to propel it at that moment.

We expected him to gag on the gas, but it seemed to have no effect on him. Nevertheless, after a few minutes, he lost interest in our camping equipment and ambled down the beach and eventually out of sight.

Now that this beast was gone, I noticed the sun was getting lower in the sky. it was getting late. When the adrenalin is pumping, time flies. We had arranged for Tim to pick us up at 5 p.m., and it was already 5:30 p.m. Now there was nothing to do but wait.

By 6 p.m. it was obvious there was some sort of problem. We hated the idea of spending the night with these grizzlies. We made plans to bushwhack back to the promontory on the mountainside with the one remaining tent. The scary part was knowing we would need to trudge through the thick willows and tall grass and all those bears to reach the supposed safety of the hillside promontory.

We had just started hiking when way off in the distance we heard the faint hum of a small plane. We hoped it was Tim. We waited, and it was him. He landed the Cessna on the lake and taxied right up to us. I noticed the plane looked different somehow.

Tim explained that he was late picking us up because he had to search for two caribou hunters who had been lost for a couple of days. Fortunately he had found them and returned them safely to Iliamna. Then Tim had returned for us in a different airplane. We were happy to learn it had two good magnetos.

Tim asked, "Did you come up with a name for this creek?" I said, "We sure did."

He asked what it was.

I told him, "Nightmare Creek."

He gave me a puzzled look. I just said, "Trust me, you don't want to spend the night here."

Cocos
ISLAND
N
W
E
S

SEARCHING FOR THE TREASURE OF LIMA, 1995

THE DINING ROOM at home was bustling with activity. Carol was serving dinner. Hatch and Mary Nell Walker were there. Hatch was well into his eighties. They had helped Carol with house cleaning chores for many years, and they were like family.

Dooley and I were already at the table. I tapped my knife against a glass as everyone gathered around, and I said formally, "I have an important announcement to make."

They were all was curious and excited to hear what I had to say, because I didn't make formal announcements often. I turned to Dooley, who would have been about twelve years old at the time. "Dooley, I am going to take you on an exciting expedition to a place that is very remote."

Everyone leaned in to hear this. I was feeling important. I said, "Dooley, I am taking you to a place that is so remote, I can assure you that no one at your school has ever have been there. In fact, it is so off the beaten path, no other kid at any school in Knoxville has ever been there. I'll even say with confidence that so isolated is this place, no one in Knoxville has been there—and perhaps no one in the whole state of Tennessee."

Admittedly I was overdoing it. Nevertheless, it was great to feel the suspense in the room. This announcement was going exactly as I had planned, with each of them eager for this mystery place to be revealed. I was basking in the attention. Every eye was on me. They all wanted to know the location of this secret destination.

"I'm taking you to one of the world's largest uninhabited islands. It's called Cocos Island, Isla del Coco, to be precise. It's way out in the Pacific, off Costa Rica, and all by itself in the ocean, and I can assure you that no one we know has ever even seen it."

Then Hatch Walker said the most unexpected thing I ever could have imagined. He said, "Oh, I've been there."

"What?" I said in astonishment.

They all looked at Hatch, and I was in a state of disbelief. Then he added the most dumbfounding statement I had ever heard: "Yep. I took the president of the United States fishing there."

"Oh, come on, Hatch. That's impossible," I said.

"Yep," said Hatch. "It was 1940. I was in the Navy and assigned to a destroyer, part of our fleet well off the coast of Central America guarding the Panama Canal just before the outbreak of World War II. We got a call one day from the admiral, and he said President Roosevelt wanted to see Cocos Island and even go fishing there and that he would be flying in the next morning."

Hatch continued, "I ran one of the landing crafts, and they assigned me to take the president fishing. I installed eye bolts on the deck of my boat," said Hatch, "so I could attach the president's wheelchair and stabilize it in the rough seas.

"When the president arrived by seaplane, we stowed his gear on the destroyer, and I took him out fishing. He was such a nice man. We had a great afternoon together. Best time of my life, and we got along splendidly. President Roosevelt even hooked and landed a sailfish. I've never seen a man so happy."

By this time, everyone in the room was enthralled in Hatch's incredible story and had totally forgotten my proclamation that no one in Knoxville had ever been to Cocos Island.

Hatch was a member of our country's greatest generation. He had fought throughout the war in the Pacific and was on one of the destroyers the day Japan bombed Pearl Harbor. You sometimes had to pull the stories out of people like Hatch—the ones who had been there and lived through it—and when you did, the stories were profound.

Cocos Island is, in fact, one of the world's largest uninhabited islands. And even though Hatch Walker took the president of the United States fishing there, it's safe to say that few in that year of 1995 had ever heard of Cocos or could even find it on a map.

Unlike other islands that are part of an island chain, Cocos sits all by itself—approximately four hundred miles out in the Pacific off Costa Rica. No wonder pirates had found it such a perfect hideout.

The best-known legend of Cocos is *The Treasure of Lima*. In 1820, with an enemy army approaching, the viceroy of the city entrusted the treasure of Lima to British captain Thompson for safekeeping, but Thompson sailed to Cocos and buried the treasure instead.

Shortly afterward, he was captured. All the crew except Thompson and his first mate were executed for piracy. The two said they would show the Spaniards where they had hidden the treasure. But after landing on Cocos, they escaped into the jungle and were never seen again. Stories like these made Isla del Cocos the inspiration for Robert Louis Stevenson's book *Treasure Island*.

Our trip came soon enough, a four-hour flight from Atlanta to San Jose and another, much shorter flight from San Jose to Punta Arenas. Here Dooley and I met my friends, Trevor Cockle and Randy Baker, from the famous sport fishing boat the *Hooker*, and boarded for the long voyage to Cocos.

While in Punta Arenas, I had purchased a pocket-size guidebook about Cocos. We left Punta Arenas in the early afternoon, and my plan was to read it on the long trip by boat to the island.

I've long since misplaced the little guidebook, but I recall that the most intriguing part of the book was one of the middle chapters entitled "The White Dove of Paradise." According to the book, and these are the approximate words from my memory of it, "When someone new first sets foot on Cocos Island and enters the jungle, the visitor is greeted by the white dove of paradise. And if the visitor stands perfectly still, the bird will circle down out of the trees, hover in front of the visitor's face like a hummingbird, slowly circle the visitor's head, and then fly off into the jungle never to be seen by that visitor again."

I didn't know whether to believe that story or not, but I intended to find out if it was true.

In addition to the guidebook, I had brought with me a copy of Zane Grey's great classic, *Tales of Fishing Virgin Seas,* in which he described his exploratory expedition to Cocos. The book was first published in 1925, but I would soon find out that Zane Grey's experience exploring Cocos and the encounter that Dooley and I would soon have there were eerily similar.

I read most of the long night with little sleep due to the bad weather and rough sea. Once or twice I put the books down and tried to get some sleep, but I was too keyed up about the trip, and sleep never came. I tossed and turned and thought the morning would never come. But finally it did, dawning cool, windy, and gray. Then, just as it was on Zane Grey's trip so many years before, through brief breaks in the dark clouds and rough seas we got our first look at the island, mysterious in the spitting rain and ominous with huge waves pounding angrily upon its shores.

Even after reading Zane Grey's book, the mountains appeared taller than I had imagined, and the whole island seemed dark and unwelcoming against the stormy sky. As we approached from afar, there was a feeling of great mystery. The island seemed lost and all alone in the vast Pacific—something apart from the world Dooley and I had come from and bearing traces from too far back in time for us to ever understand.

We marveled at the rugged cliffs and vast wild jungle, shrouded in mist and rain, each of us imagining what it was like for the pirate ships that visited here. Time and again I caught myself scanning the beaches and mist-covered valleys, wondering where I would have buried the treasure if I had been Captain Thompson. I quickly decided that the terrain was so wild and the jungle so infinite and thick that the Treasure of Lima might remain hidden for the rest of eternity.

Zane Grey had mentioned the staggering number of seabirds at Cocos, so I was prepared to see a lot of birds. At the age of fifty-three I had spent years on the ocean, but I had never witnessed anything like this.

Boobies, terns, and gulls—it seemed like millions of them, and they darted and sailed across the rough Pacific and filled the sky as far as Dooley and I could see. They came like an outpouring from every nook and cranny on the island, soaring high and skimming the tops of the waves—swarms of seabirds with piercing screams, diving straight down at breakneck speed into the sea after prey. There were also giant man-of-war birds (*tijeretas*), screeching in a mean-spirited way from just above our outriggers and diving toward the *Hooker*. In a very real way, they seemed like something almost prehistoric, screaming viciously as if they were the appointed guardians of this strange island.

In fact, the island itself was the inspiration for Michael Crichton's great book *Jurassic Park,* and the opening scenes of Spielberg's movie were shot here. As we approached the island, it seemed almost reasonable that there could be prehistoric creatures in its vast mountain jungles.

Two mountain ranges rose from the sea, shrouded in mist, their highest peaks hidden by dark clouds. As we ventured closer, we realized the sides of these steep mountains and rugged valleys were lush, and the color of it seemed a richer shade of green than even the rain forests on the Costa Rican mainland.

Then we saw the waterfalls tumbling downward—torrents of white water that seemed to pour from the soul of the earth and float down from the high mountains and cliffs in slow motion, some of them freefalling straight into the ocean.

This island was idyllic, but as the days of our trip flew by, we often had this vague feeling of something eerie—a haunting intuition that there were secrets here on Cocos not intended for two guys from East Tennessee to know or understand.

I left the *Hooker* early one morning and went ashore alone in the small panga. The moment I stepped onto the island, I felt the grip of solitude—and some feeling of unbelonging. I crossed the sandy beach, captured by the mystic. I passed beneath the coconut trees, all of them bowed and warped by the relentless ocean winds. Beyond them was thick elephant grass, five or six feet in height, waving silently in the morning breeze.

I made my way through the tall grass and entered the great rain forest. It seemed alive and strange to me. The trees were different than any I had ever seen, stretching high above and blocking the sky, shielding the light from the sun. Vines of every dimension stretched downward as in some old Tarzan movie.

I stopped and listened, and the sounds were ethereal. I had visited Costa Rican rainforests on the mainland, but here on Cocos there were sounds I had never heard before. I stood there transfixed, motionless and mesmerized.

It was in that moment that something white flickered in the tallest of the trees far above me. I caught a glimpse of something sparkling as it sailed and spiraled downward, coming closer until it was circling just above my head. It was a majestic white bird, almost like a dove, only larger, the size of a pigeon—but pure white. It stopped directly in front of me at eye level and hovered there like a hummingbird.

How could a bird that large hover? It continued to move closer until it was about a foot and a half from my face. I stood motionless. I dared not flinch or blink lest I frighten it away. I looked directly into the eyes of this beautiful creature, and as it looked back at me, it seemed to be looking into my soul. The wind from its hovering wings was cool and constant against my face.

Soon it moved to hover by my right ear. I dared not turn my head for fear of frightening it, continuing instead to look steadily straight ahead. I could feel the blast of wind from its wings as it moved to hover behind my head, and while it was there, I had the most unlikely and profound thought—that perhaps the white dove of paradise knew where the treasure was and that maybe it was the only one who knew. The bird moved again to hover by my left ear, where it stayed even longer. I sensed what would happen next as it moved back into my sphere of vision and hovered directly in front of my face, looking deep into my eyes. How long it stayed there I can't say. I was too enthralled. It was a moment of unknown length frozen in time.

To this day, I don't know if it was a greeting or an inspection. I guess I'll never know. But in a heartbeat, the magnificent white bird wheeled up and away, spiraling higher and higher above my head—up and up and through the tallest of the trees far above and out of sight. And I never saw the bird again, or one like it.

It was as if I had ventured into some liminal dream world, a mysterious realm between reality and enigma—alluring yet somehow foreboding in equal measure. An island bypassed by man and lost somewhere in time.

I felt the breeze from the sea, and it was good to be back in the panga as I steered it away from the shore. The boat was anchored in the lee of the island in fifty feet of water. The ocean in this half bay was crystal clear. By day, Dooley and I saw sand sharks asleep on the pebble ocean floor. It was a peaceful place to swim and snorkel by day, but in the darkness, it was otherworldly and perilous. The sharks owned the night.

After dinner we turned on the spreader lights from the bridge of the *Hooker*. In the darkness, what Dooley and I saw in the water was unnerving: sharks by the hundreds and thousands, as far as the lights would illuminate. So many sharks, all side to side and back to belly, and so thickly stacked that if they had held still, I could have walked on them.

I had heard some say that Cocos had the highest concentration of sharks in the world. I never knew for sure, but at that moment, I believed it was true. As Dooley and I stood there on the *Hooker* with the illumination from the spotlights above, we saw threshers, whitetips, silkies, blacktips, hammerheads, and others I could not identify. We did not see whites and tiger sharks, but they were likely lurking somewhere in the depths or nearby.

We amused ourselves by feeding them the remains of the wahoos and dorados we had caught and cleaned for dinner. The instant the blood hit the water, it triggered chaos. I had seen sharks feed before, but never this many. Like a frenzied horde of starved wolves, they fed viciously, often biting each other and churning the water into a bloody froth. It was terrifying.

Other nights were even more interesting, with an event we called shark skiing. It started when Randy Baker found an old ski rope in a storage locker on the mothership. At the end of that rope, we tied a slipknot to hold the bait.

Each of us then applied suntan lotion to the bottoms of our bare feet to slick them down. The skier then grabbed the wooden handle of the ski rope while one of us tossed the other end of the rope, with the fish remains on it, to the sharks. We tried to target one of the larger sharks to get a good, fast ride. At this point the shark would take off with the bait, pulling the skier barefoot across the long, wet, and slick deck of the *Hooker*. You had to stay low to be sure the transom of the boat would stop you. This became our go-to nighttime activity, and more than a few bets were placed on who got the fastest ride.

By day Dooley and I fly fished for striped marlin on the drop-offs a few miles offshore. It was like fly fishing off the coastline of Jurassic Park—a fitting thought, as marlin had ruled the oceans for millions of years.

At the age of fifty-three I had no way of knowing that I would return to Cocos three more times in my life and set two fly rod world records here on striped marlin, barely missing a third record by only one pound. For me, those adventures were yet to come. But here and now, on this first expedition to Cocos with Dooley, it was about something more than fly fishing. It was discovering and experiencing this remote place with my son.

It was almost impossible to visit this deserted island and not consider what it would be like to be stranded here. Dooley and I discussed it often and at great length. We debated whether we could survive here on our own or for how long. It was entertaining to fantasize about such a thing, and each of those conversations always ended with a confirmation that being stranded here might get old fast. Not to mention the mystery and loneliness of this place. We decided that Cocos was a most intriguing destination to visit for a while, but only if there was a way back home.

I remember our last day in Cocos. A light rain was falling as we pulled anchor and set a course for the mainland. From the open bridge of the *Hooker,* Dooley and I stood together and watched the lost world of Cocos as it disappeared on the vast horizon.

With our faces to the wind, we felt the freshness of the rain mixed with the aroma of salt from the sea. Above us were the *tijeretas,* screeching and diving at us, undoubtedly convinced they were driving us away from Cocos. I thought about the white dove and the mystery of this place—about the pirates and Captain Thompson and the treasure that was never found. As Dooley and I stood together, facing the oncoming wind, it struck me how often we humans search for treasure, sometimes throughout our lives, in every place except the very place it is. And it was in that moment that I realized where the treasure of Cocos was.

I had figured it out, and I was sure of it. The treasure of Cocos was not on the island. Not at all.

The great treasure of Cocos was at my side.

PART SEVEN

Pushing Boundaries

CABO VERDE
(CAPE VERDE ISLANDS)
SANTO ANTAO
SANTA LUZIA
SAO VICENTE
SAO NICOLAU
AFRICA
North Atlantic Ocean
BOA VISTA
SANTIAGO
MAIO
FOGO
N
W
E
S

THE HUNDRED-YEAR MARLIN, 1998

IT WAS A LONG nighttime flight over the Atlantic, and I watched a movie that I had seen before. That's when I reached in my pocket for the little Ambien pill. It was in the same pocket that held the blue marlin coin, the one my dad had brought back from Bimini so many years ago.

The coin was not only a reminder of him, but also of the good things I loved about fly fishing for marlin—the scent of salt and seaweed in the morning breeze, rays of light shimmering through prisms of cobalt blue, the sound of seabirds, ferocious strikes, wild and savage runs, frenzied, twisting jumps, raw power, and the thrill of the chase. Butterflies in the stomach for the fight to come and not knowing how long it would last. The mystery in the depths below with no clue as to how big the next fish would be. Maybe too large for a fly rod, which brings me to another aspect of fly fishing for blue marlin—the part that's brutal. The fight on a fly rod is long and hard. It takes something from inside of you, from the very core of your being, and in the aftermath, you are something less than what you were before.

I had plenty of time to contemplate these thoughts. I was on a Boeing 747 from New York City to Johannesburg, South Africa. The only problem was that the aircraft didn't have enough fuel to get there. There was a planned fuel stop in a small African island chain nation 385 miles off the coast.

The plane landed in Santiago, the largest of Cape Verde's ten islands, for twenty minutes of refueling. There were 414 passengers on the flight, and 413 of them stayed aboard the plane during refueling and went on toward Johannesburg. Only one got off in Cape Verde. Naturally, that was me.

Even stepping off the plane, it was obvious at once how poor this island country was. The hillsides were steep volcanic rock. The mountains on some of these islands rose from the sea to a height of nine thousand feet or more—mostly barren volcanic rock and dirt.

At the age of fifty-six, I wasn't here to sightsee. My goal of catching large fish on a fly rod had by now morphed into something more. No longer was my dream to simply catch big fish. Now the goal was more fly rod world records on the strongest, fastest fish in the ocean: blue marlin. Here in Africa, I hoped to do it on 16-pound test tippet, and if successful, to try for another record on 12-pound test.

The best blue marlin fishing in the world was rumored to be here in this very place. I wondered if the marlin liked it here because it was seldom fished. More likely, it was a feeding ground where strong currents of the deep Atlantic collided with the volcanic underwater ledges of Cape Verde and attracted the baitfish the marlin fed on.

I would be fishing with Captain Trevor Cockle and mates Randy Baker and Ronnie Fields, friends who I had fished with before. I would also be fishing on my favorite sport fishing boat, the *Hooker,* a forty-seven-foot G&S. Naturally, my confidence was high.

But after a couple of days, we weren't seeing many fish. So on March 19, we moved the boat from Santiago to the island of Sao Nicolau. The morning of March 20 dawned extremely windy and cold. The seas were high and rough.

Around 10 a.m. we eased the *Hooker* into the lee of the island where the water was considerably calmer. Immediately I caught a rather large Atlantic spearfish that weighed in at sixty-one pounds—a new fly rod world record. Amazingly, the very next day I set the Atlantic blue marlin record on 16-pound-test tippet.

The expedition to the Cape Verde islands had already been a success, with two world records and three days left to fish. The pressure eased, and my confidence was high. I felt ready to tackle anything with my fly rod.

The next morning, March 21, dawned glorious. The Cape Verde sunrise was right on time as Trevor cranked up the two Cummins 903 diesel engines on the *Hooker*. The wind had calmed. It was the kind of day that fills you with hope and expectation.

We made the short run back to the area where we had hooked the world record marlin the previous day. Seabirds were everywhere, huge rafts of them floating on the ocean like a thousand buoys, with hundreds more flying high and diving at the baitfish—always a good sign.

I decided to devote the last three days to fishing with 12-pound test tippet. At that point in time, no one had ever landed a blue marlin on a fly rod with 12-pound test tippet. I wanted to be the first. I also rigged a second fly rod with 20-pound test tippet in case we raised a fish that was way too large for 12-pound test.

The teaser lures were out, and once again we were fishing. We raised two nicely sized sailfish in the first forty-five minutes, and I landed both on 12-pound-test. Trevor headed the *Hooker* back in the direction where both sails had hit.

Suddenly there was a massive swirl under the left long teaser. Ronnie grabbed the teaser rod as something enormous crashed the teaser a second and third time, gigantic explosions of water and spray caused by some monster yet unseen. Out of the melee, a mammoth dorsal fin of a marlin appeared. Ronnie was reeling the teaser to the boat at near rocket speed to keep it from the jaws of this aggressive beast.

The marlin was big—one of the largest I had ever seen while fly fishing, and way too big for fly tackle. But what the heck: I'm a fly fisherman, and I had to feel the power and fury of this fish even if only for a few seconds.

I felt a tightness in my stomach, but there was no time to get overly nervous. Instinctively I knew this fish was way too large for any fly rod, particularly one with a 12-pound test tippet, but I wasn't going to let this opportunity pass. I grabbed the rod with the 20-pound test tippet. Trevor kicked the *Hooker* out of gear. Ronnie desperately yanked the teaser lure into the boat as this monster blue marlin lunged for it. I flipped the fly to him almost out of instinct. Deep inside, I knew this beast was way too large, but there was no stopping now. The fly landed close to him and to his left. The leviathan simply turned and ate my fly, heading away from the boat.

I was fortunate not to break him off on the strike, especially at such close range where there was no stretch in the line. I quickly backed off on the drag to zero, expecting the frenzied fast run that usually breaks my leader tippet. I figured with the size of this huge marlin, I'd be lucky to hold onto him for more than a minute or two at best.

I heard Trevor yell from the bridge that he was well over five hundred pounds. The marlin took off like a freight train, and his jumps were long, titanic leaps. Each time he jumped it was like the soul of the Atlantic surged within him, and each leap was followed by a body slam, shattering the surface of the ocean. These weren't ordinary jumps; he was much too large for that. On each of these vaulting leaps, he exploded back into the ocean like a dump truck dropped from a low-flying airplane. And always when he jumped, I bowed the rod to him, creating slack so the weight of the fish as he accelerated into the air and on his downward momentum would not break the tippet.

More than once, he ran a mile or more just under the surface with the tip of his tail clearly visible. He was showing me that he was large and in charge with his brute force. The *Hooker* was running nearly full tilt in reverse, and I managed to stay within 250 yards of him, expecting him to go deep but hoping he would not. There wasn't a ghost of a chance of leveraging a fish of this size from the depths of the Atlantic on a fly rod.

Occasionally he would go down ten feet or so, but immediately he would be right back on top doing those same vaulting leaps, moving away from the boat and its angler from East Tennessee in desperate pursuit.

I looked at my watch. It was 10:45 a.m. I couldn't believe it. I had been fighting this giant for an hour, and unbelievably, I was still connected to him. A fish this big on a fly rod should have broken me off long ago.

I took the reel off total freespool and added just a touch of drag—just enough so I no longer had to palm the reel to keep it from overrunning. For the next three and a half hours, I fought the fish over literally miles of ocean. He must have jumped fifty times or more, but with the passing of each hour, the jumps became more like lunges, and I added a little drag on the reel.

A blue marlin of between five and six hundred pounds on a fly rod should have already snapped the tippet, and I was mystified that the light leader remained intact. I was even beginning to gain a modicum of confidence. After four and a half hours it was dawning on me that the beast and I were taking this fight to a place that fly fishing had never been and was never meant to be, and I let myself wonder what it would be like to get this monster blue marlin close to the boat. I had no idea what mayhem would happen then, but I knew it wouldn't be pretty. It was only a thought, and something to worry about later if, by some miracle, the fight got that far.

For now, I just needed to guard against breaking the thin tippet on one of his vaulting leaps and lunges sometimes as close as fifty yards. I needed total focus.

At five hours into the fight, the fish and I were tiring, but I saw him up close, and I was reinvigorated. The adrenaline was flowing. I put just a little more drag on the reel and dug deep for the strength to apply more pressure over the back of the fish.

At 3:45 p.m., the sixth hour of the fight, the muscles in my arms were about to burst. But the sight of this magnificent marlin energized me, and to see it so large, and so close to me, gave me the extra energy I needed. I realized that this was a fish larger than any man with a fly rod had ever seen at six hours into a fight.

The fish of a lifetime—of several lifetimes, actually—was right there plain as day, like a submarine near the surface and only forty-five feet away. Randy and Ronnie had their gaffs out, and both guys were tied by ropes around their waists to the deck of the *Hooker* to keep them from being yanked into the ocean if they got a gaff into the fish.

Trembling from exhaustion, I had lost nearly all feeling in my left shoulder and throughout my back and legs, but I applied more pressure. Yet no matter how hard I leaned back against the rod, it was a complete stalemate—a Cape Verde standoff. The monster marlin could not take more line out, and no matter what I did, I gained no line on him. But after another thirty minutes, he tired just a bit, and as hard as it was for me to believe, this enormous marlin had come closer to the boat—near to gaffing range. The adrenaline was pounding in my chest as Ronnie and Randy hit the fish simultaneously with the two gaffs. There was a massive eruption, an explosion of sea water to the height of the bridge, and both mates were yanked violently halfway over the port-side gunnel. They both would have been in the ocean except for the ropes holding them in the boat. The great fish pulled both gaffs out of their hands and was now headed down-sea on yet another run. Line was burning off my reel, only this time, the marlin was pulling not only against me but also against the weight of the two steel gaffs.

For the next hour, he swam fast on the surface, often lunging half out of the water. We were able to stay within fifty or seventy-five yards of him, and often, even closer—close enough, in fact, that the mates had my two spare gaffs out. In the seventh hour of the fight, Trevor tried to back the *Hooker* faster in hopes we could get a gaff shot off the transom. But every time we got close, one of the gaffs already in the marlin near his tail would hit one of the props under the boat. It happened three or four times and made a loud clanging sound. It worried us, as we obviously needed the props for many reasons.

After several tries to no avail, we changed strategy. Trevor moved the *Hooker* off to the side of the marlin so we were backing down in reverse, parallel to him off our port side. Meanwhile, Randy and Ronnie had the two new gaffs at the ready. They retied themselves to the boat, waiting for another chance.

The heat of the sun and the constant pull of the fish against the rod were wearing on me. I had been standing up barefoot and fighting this beast for well over seven hours. My feet were numb, and I couldn't feel the deck. My legs were shaking, and my arms felt like lead. For a moment, I wondered what I was trying to prove. I had already set five world records in my life, and in a way, the extreme torture of this fight made no sense. Then I erased that thought from my mind, and instead, I focused on the enormity of this fish—the enormity of catching the largest fish in the history of the world on a fly rod—and I was determined to see if it could be done.

Now with this giant so close, adrenaline had me pumped. I realized fully that this impossible catch was now possible. The marlin was within twenty feet of the boat and near the surface, the tip of his tail always visible, gliding and winding back and forth above the glare on the surface of the Atlantic. He swerved away, and as the boat turned, and with the late evening sun behind me and his back now out of the water, I had the clearest look at him since the fight began. I couldn't believe what I was seeing.

He was twelve feet long or more, every stripe brightly radiating and contrasting with the purple and blue of his body. His eye appeared to me to be as big as a saucer. I put all the pressure on the rod that the tippet would take. It seemed as if I moved him—or was it just wishful thinking? I needed him much closer, but this fish was more powerful and relentless than any fish I had ever fought or imagined.

Then, at last, there was a moment when he seemed to weaken just a bit, and I leveraged the great marlin within nine or ten feet off the port side. The two eight-foot gaffs in the hands of Ronnie and Randy were only inches away from this beast. I was applying all the pressure the line would take to move him even closer for gaffing, and I was close to the point where the tippet would break.

Suddenly, he rolled up on the surface and lay there nearly spent. I didn't want to risk breaking the tippet now that the great fish was totally exhausted and ours for the taking. I backed off on the pressure just slightly, knowing I had won. Why risk breaking my tippet now with the fish so tired and unable to fight? I was confident knowing that Trevor could simply move the *Hooker* over a few inches so Randy and Ronnie could get their two new gaffs in the fish, which was now lying there on the surface totally spent.

And then he started sinking.

What none of us had realized was that the weight of the two steel gaffs already hanging off the underside of this giant marlin would make him sink once he couldn't swim anymore, and once I backed off the pressure. And in that second, when we could have gaffed the great marlin so easily, he sank downward into the dark blue depths of the Atlantic.

I tried to lift him, of course—but a fish of more than five hundred pounds plus the weight of two steel gaffs, combined with the current of the ocean, made it impossible on a fly rod. I watched him sink down and down and down, through the prisms of blue, and finally out of sight. And once he was down a hundred yards in the deep below us, the tippet finally broke.

It was a battle of eight hours, an entire day. It was a fish too big for any fly rod. Yet, incredibly, I was literally within an inch of success. This impossible marlin would have shattered my own Atlantic blue marlin fly rod record on 20-pound test tippet by more than three hundred pounds.

All records are made to be broken. It will happen to all of mine as well. Today, in early 2022, as I finish writing the stories in this book, my original blue marlin record set in Venezuela in 1994 has stood for more than twenty-eight years. I think I can safely say that had I landed this beast in Cape Verde, which was in the five-to-six-hundred-pound range, it would have been a fly rod record that might have lasted for a hundred years or more—and to this day, I refer to him as "the hundred-year marlin."

Ernest Hemingway reportedly said, "Big-game fishing should be a fair contest between fish and fisherman." If that is true, I more than did my part by fighting this brute on a fly rod for eight hours. Throughout my life, I remember the toil of this battle, and this lost fish, more than any fish I ever landed. Often when I lie in bed at night and can't sleep, I relive this day and this fight, over and over. I try to rethink it. To make it end differently. But it always ends the same.

We were tired that evening, all of us, and we turned in early. I closed my eyes and listened to the gentle waves lapping against the boat, and I wondered if I would ever again see an ocean with so many blue marlin. I thought to myself that if I was ever to land one on 12-pound test tippet, this would be the place.

Then I remembered the immense size of these marlin, and I wondered if I could ever land a marlin the size we were seeing with only 12-pound test tippet—and how long the fight might last, and if there was a fly rod angler anywhere that could do it.

THE FIGHT OF A LIFE, 1998

THE MORNING OF March 28 dawned fresh and inviting—a good day for a chase with just a hint of cool breeze in the Cape Verde sun. We rode a friendly sea that promised adventure. The ocean was blue, the sky was blue, and way out there as far as I could see on the horizon all of that blue seemed to come together as one, and I knew that beyond that there was even more blue. I thought about the immensity of the sea—the place where all the rivers go, where life itself began, where the fish were large, where the flies that I had tied were oversized, and where everything seemed magnified.

This whole expedition and the dream I pursued—it seemed as if I was on a quest to see how far I could push the sport of fly fishing in saltwater, and I was determined to catch one of these big blue marlin on 12-pound test tippet.

Somehow the eight-hour fight with the big marlin the day before had changed me. It was as if I had bridged a gap between skill set and belief. Now I had the confidence that I could not only fight big fish, but that I could go the distance with any marlin I hooked on a fly rod, and I was ready for battle.

At 10:00 a.m., something large smashed the left long teaser. The huge sword of a marlin sliced through the surface of the Atlantic, busting and slashing the teaser lure repeatedly. I stood barefoot in the starboard corner of the transom, the fly rod clutched tightly in my right hand, alert to see what would happen next. But there was nothing. The marlin had vanished.

Then, in a heartbeat, the marlin reappeared behind the right long teaser, stalking it like a stealth bomber as Randy teased him closer and closer to the boat. The marlin was coming toward me with a swagger, like an NFL linebacker in a blue jersey, intent on smashing anything that got in his way. It was almost frightening.

I would need my back-cast to be low underneath the right outrigger so as not to hang up there and lose my chance at this big fish. There was no time to think. The cast was fast, mostly by muscle memory as the boat came out of gear and the teaser lure ripped from the water and whisked into the boat.

My fly landed in the clear blue water to the right of the propwash. The fish swam right by it because he was keyed in on the teaser lure. But when the lure left the water, the big marlin swirled around to find it. The only thing he saw was my blue tube fly, which he attacked viciously with intent to kill.

There was no way to not hook this fish. He rushed the fly so violently and quickly that he hooked himself. My only thought was how crucial it was to not let him break my thin 12-pound test tippet.

As I backed off on the drag, he jumped close to the boat, and I could see that he was big. I heard Trevor shout out from the bridge, "Two twenty-five," meaning 225 pounds—definitely bigger than I had hoped for on 12-pound test tippet. But after the events of the preceding day, nothing seemed impossible on my fly rod.

The fish accelerated straight away for three hundred yards, never once visible on the surface of the ocean, and then he dove deep—deeper than I had ever seen a marlin go.

And soon the fight became a slugfest with the depth and the current in his favor. I applied a lot of pressure on him, and occasionally I gained some line, but always, he would take it back, and a lot more. We maneuvered around in front of him to get a better angle with the boat, but he just went deeper or changed direction. For hours he swam deep and steady, and he had the advantage as I struggled on and on with no success. It was now high noon. The marlin and I were locked in a battle of duration. I kept constant pressure on him as the hours crept slowly by. I checked my watch. It was 3 p.m, and it was hot in the African sun—the kind of heat that will drain you.

As the hours went by, I hung on to him, but it was grueling. The pain was there again, throughout my arms, shoulders, and legs. It was a war of will and a battle to see who would last, and through it all there was constant strain on my entire body.

At 5 p.m., after seven grueling hours, he finally surfaced. He jumped once, and jumped again, and then he went deep. I knew he wasn't the slightest bit tired, but I was, and with his strategy of staying deep, he clearly had the upper hand.

I had now fought this fish most of the day, and with darkness approaching, there would likely be a long fight in front of me. I settled into that zone where I knew it was not going to be a sprint but a grueling marathon. The danger was putting too much pressure on the fish and breaking the fragile 12-pound test tippet. Yet I would need to keep constant pressure on him to force him to come up and fight me on the surface.

The African sun set sadly on the horizon, perhaps a foreshadowing of the night to come. Under darkness sharks feed aggressively. The marlin know that all too well, and I hoped that knowledge would make him nervous and spur him to make a move. But it did not, and he swam on and on, deep and unwavering, always on the move.

Darkness swallowed the ocean as heavy cloud cover moved in. No stars were visible. At 8 p.m. I felt the wind begin to build. The fish was still very deep, and on a course running up-sea, which meant we were backing down on him into heavier waves. As the wind picked up, each of those waves began to break over the transom of the boat, drenching me.

It's an eerie feeling being way out in the Atlantic on a windy and dark night in rough seas and not being able to see anything, not having a bearing on anything solid, and fighting a big fish and high waves—not knowing how much rougher the sea will be or how long the fight will last. It helps to have a little grit and a streak of stubbornness in your makeup when you are in the tenth hour of the fight, the sky is dark, the seawater crashing over you is cold, you feel your entire body begin to shiver, and you are chasing the blue marlin you've dreamed of on 12-pound test.

You need total concentration to fight a marlin on a fly rod, but after ten hours of strain on the body, your mind begins to wander. For a moment I was a kid again, remembering the time my dad had taken me to Bimini and I had seen Hemingway and overheard stories of great battles with marlin and tuna. The strength of this fish reminded me of his book *The Old Man and the Sea,* and for a moment or two I thought that perhaps this fight was like that. Then I doubled down in my determination, putting myself in a mindset of total commitment and erasing all but the most positive thoughts.

Silently I said to myself, "There is nothing in the world that exists but to land this fish. My only focus will be the marlin, and I will think of nothing else. No matter how long the battle lasts, sooner or later the marlin will tire, and when it does, the advantage will be mine." It's something that happens whenever I fight a big fish: a moment in time during the fight where I become the predator, like a wolf tracking an elk or buffalo—running with it, on and on across the hills and sagebrush prairie, until the prey can run no more. It's an advantage I possess—an inordinate dose of persistence—a wolf with a fly rod.

By now it was 10 p.m. and the battle with the marlin was at the twelfth hour. The wind was heavy, and the waves breaking over the transom were cold. I was shivering, wearing only shorts and a T-shirt. I knew that eventually my shivering would cause me to make a mistake. I needed warmth to even up the fight with this fish, or he would win.

Randy brought me my rain suit. It was tricky getting into it while still holding the fly rod, one leg and arm at a time, with the boat lurching violently and each big wave smashing over me. But once I had it on I warmed up a bit, and after a while the shivering was gone.

Around midnight the wind let up some, and the seas were no longer breaking over the back of the transom. Suddenly the line came to life. It was violent and hard, and I thought maybe he was finally going to make his move to the surface, but the marlin simply changed direction. He stayed deep and began running down-sea. I wondered if I had forced him to change direction. Was it a sign he was getting tired?

I love to fight a fish going down-sea. It's a smoother ride. I had now been standing up barefoot on the slick deck of a lurching boat in surging seas for fourteen hours. I was taking fly fishing for billfish to a place it had never been. I relished and fed off that thought, gaining more and more confidence from the smoother down-sea ride. I felt more secure in my foothold on the wet deck of the *Hooker*, and I focused even more.

It was 1 a.m. when Randy said, "Do you realize that you've had this fish on for fifteen hours with nothing to eat or drink?" I had been so focused on fighting the fish, the thought hadn't even occurred to me. Randy brought me a roast beef sandwich and a bottled Coke. I continued putting pressure on the fish with my left hand on the rod and used the right for the sandwich and Coke, which disappeared quickly. I realized how dehydrated and hungry I had been. Now the combination of the food and the down-sea ride energized me, and all my instincts were zeroed in on fighting the marlin.

Hour after hour and mile after mile in the dark Atlantic the fight endured, and I hoped that he might come to the surface where I might have an advantage. But as the night wore on, my legs were tiring, nearly always trembling, and the balls of my feet felt as if I had walked a thousand miles, and there was pain in every part of me. And all night long the fish stayed deep, conserving his energy in this contest of who would quit first.

In a war of attrition, the opponent with the best strategy wins. By now, I had fought this marlin for eighteen hours, and his strategy never wavered. He stayed deep, showing me his power, making it clear that he was in charge, and no matter what I did, he held the advantage.

It was the dark side of dawn when I looked at my watch. I had now fought the marlin for nineteen hours. My back was weak, and I had lost almost all feeling in my arms and legs. Although I am not a boxer, I thought maybe it was like the fifteenth round of a heavyweight fight, and with that thought, I dug deep, and I stayed with him.

As the first ray of light dawned over the eastern horizon and signaled day two of the fight, the fish suddenly rushed upward toward the surface. With an amazing burst of strength and speed, he tail walked the ocean for five hundred yards. I backed off to zero on the drag and palmed the reel ever so lightly, amazed and in awe of his speed and energy, and surprised that my frail 12-pound test tippet had lasted so long.

My instincts were telling me, "This is where I wear him down and land him." But at the end of that amazing run, the marlin dove deep yet again—though not as deep as he had been all night. I wondered if the long fight was beginning to wear on him.

Apparently not, because as the sun rose higher in the sky, he dove deeper into the depths and took out more and more line. I applied more pressure than I had throughout the night, but he went even deeper. Mentally I was still strong and into the fight, but the ache throughout my body was constant and intense, and I had lost all feeling in my feet. All the while, the marlin was like a great athlete.

I marveled at his strength.

Then, at 8:00 a.m., twenty-two hours into the fight, with the sun well up and the seas calm, he made another move. Even deep in the ocean, I could feel the sudden violence and then the startling speed. Line started screaming off my reel as he came streaking upward, out of the depths and through the surface of the Atlantic and into the African sky, the highest I ever imagined a fish could go—a sign perhaps that he would now try to beat me on the surface. Sure enough, for the next half hour he put on a nonstop show. He high-jumped over long, smooth waves and crashed through the tops of others, always visible; always in sight.

Always powerful, like he had this plan all along, to wear me out with the brutal fight all day yesterday and last night and then break me off today once I was tired.

I wanted to fight up close, but his speed was amazing—and I could not keep up. He seemed revitalized with ever-increasing velocity and power as he took out more and more line. It was impossible to imagine where he found the strength for these wild and crazed runs on the surface after fighting me for two days. To me the source of that strength was a mystery, and worthy of my respect.

For another thirty minutes, I fought to keep him from taking out more line, shocked and stupefied by his raw power and stamina. I never even got close to him, and he continued to move farther and farther away.

Then, suddenly, he showed the first sign that he might be weakening. He slowed, and at last I gained line. Then unexpectedly, he stopped cold and rolled up on the surface of the ocean, just lying on his side, with one of his big pectoral fins sticking straight up in the air like a lance. If we could just get to him fast enough, he would be easy to gaff. He was a long, long distance away, and Trevor was backing the *Hooker* down fast while I was reeling up slack, trying to keep up with the speed of the boat. It seemed to me like forever, but soon enough we came up close to the marlin. He rested on the surface of the calm ocean, exhausted and ready for the taking. Ronnie and Randy had my two new gaffs ready. As we moved in beside the marlin, they both pulled upward toward the fish simultaneously with the gaffs.

Then there was an explosion of seawater.

I'll never forget the next two seconds as long as I live.

When the marlin felt the steel hit his body, he made one last leap beside the boat. For a millisecond he hung high in the air above me, droplets of water exploding bright in the Cape Verde sun, and he broke my 12-pound test tippet.

Something like that takes a second to register. In the blur of that moment, you think you can turn back the clock. Then the reality of it hits you like a semi, and it smacks hard that there is no turning back the clock, not in fishing, not in life, not even for a second, and he is gone.

After nearly twenty-three hours extending across two days of fighting, the fish had won and was free. All four of us stood there together on the deck of the *Hooker*. We looked at each other in amazement, wonder, disappointment, and some other feeling I could not identify.

Blue marlin will do that to you.

None of us said a word; we simply stood there. The big blue had fought me well, better than any fish I ever fought. So well did he fight that it made me wonder if anyone was worthy of landing that fish on a fly rod—especially the angler from East Tennessee.

For a while the four of us seemed transfixed. We knew we had all just participated in something epic and that all four of us had poured our heart and soul into that fish. For the marlin it was the fight of a lifetime, and for me it was as well. It would take some time for understanding.

One thing we knew for sure: I was fishing on the edge, and I had just pushed a boundary in fly fishing—all of us had. Maybe on a fly rod I was going a bit too far, further than fly fishing was meant to go. It would take a while for it all to sink in—to fathom life lessons and gain wisdom. Better to head back in and get some sleep without overthinking it. We were all dead tired.

It was, to my knowledge, the longest billfish fight in history on a fly rod. Those final days off the coast of Africa are etched in my memory and will never leave me. They are part of me and who I am.

I have fly fished for over seven decades. But ironically, that brief juncture in time, those last three days in Cape Verde without a fish landed, is when I was at my best.

A COLD WIND, 1999

IT WAS LATE in the fall, and the north wind was sweeping strong across the gravel runway in the Native village. It had a bite to it, and the Arctic air cut through my windbreaker and fleece with a bitterness I had never felt before.

"You're late this year," said Tim.

Tim LaPorte and his wife Nancy ran the little airport at the remote village of Iliamna in southwest Alaska.

"We were beginning to think you weren't coming this year," added Nancy. "We haven't seen a fisherman since early October, only bear and caribou hunters."

Tim offered, "If you're headed into the backcountry, Scottie hasn't put his Cessna up for the winter yet. It's the only plane around here still on floats. All the bush pilots have put their floats away until next year. They can't afford to let their planes get frozen in."

John Bechler was there, too. We had known each other for years. John and I had once spent a week together on a remote outback tundra stream and the experience had been awesome. He said, "You're welcome to stay with me and fish the big river here by the Native village. You're sort of pushing your luck to be out there on your own in the backcountry this late in the year."

I thanked him but asked Tim to find Scottie and see if his Cessna could be ready to go in an hour. "I'd like to get into the lower river today, if he can fly," I said.

The radio crackled. I heard Scottie say, "Meet me at the pond in an hour."

Scott DeWitt was a bush plane pilot and a good one. He had flown me into many of the rivers and lakes in that part of Alaska. I loaded my gear on an ATV and drove to the pond where Scottie had his Cessna 206 nosed into the beach facing the north wind. It took a couple of minutes to load my gear into the airplane and another ten minutes or so for Scottie to warm the engine up and get the plane ready to fly.

The pond was small, and sometimes we would have to go around it twice during takeoff to get up enough speed to lift off. That's what Scottie did this day as well, and on the final part of the second circle, he headed north into the wind. Scottie and I had done this together so many times over the years.

It was invigorating and exciting to fly over Iliamna, to feel the exhilaration and the relief that everything was the same as when I had left the year before. And from above, there was so much to take in all at once. I found myself scanning the big lake and the shoals of the Newhalen River, searching its clear waters for trout. I surveyed the tree lines and game trails for bear and caribou.

The tundra, which is usually green with a tint of yellow, had already turned to rust with flecks of gold. The thought swept through my mind that change was coming soon. It would not stay this way for long, and soon the Arctic winds would have their way.

As the plane circled west and crossed the big river, there was nothing but the vast emptiness and the magnificent huge lake ringed all around by snowy mountain ranges. So immense was this wilderness that I sensed the utter insignificance of mankind. It was my twenty-fifth season in the Alaskan wilderness, and once again it was overwhelming. I felt invigorated.

We flew low across the little drainage coming in from the north. I saw a grizzly and two cubs running from the creek bed and into the tundra to our right front. It always surprised me how fast they were.

We approached the first major creek feeding into Lake Iliamna, and Scottie swept in low from the big lake at two hundred feet in elevation, heading north. It was the place I had fished with my dad twenty-five years before—the first backcountry stream we had ever fished together in Alaska—and for a moment it seemed as if it were only yesterday.

Scottie dipped the wing so I had a perfect view, and there, in the first pool above the lake, silhouetted against the pebble grain tailout, were rainbow trout, maybe fifty or more, and most of them big—trout so large they could be seen from an airplane. For a fly fisher, that experience is euphoric. It felt good to be back.

Scottie gained elevation and swung the Cessna westward across the endless tundra, and on and on we flew. Finally the river came into view, and so did the deserted old cabin, weathered with age and clinging desperately to the narrow spit of land between a smaller lake and Lake Iliamna. In the distance, I could see where the river emptied into the small lake. And finally, the outlet of the small lake and the long, straight stretch of river where it headed west, finally emptying into Lake Iliamna.

We flew in low and slow, heading up the river. It gave me a view of the shoal above the small lake. In the two-hundred-yard stretch from the small lake to the head of the first riffle, I could see hundreds of rainbows of all sizes, many of them even bigger than the ones I had seen earlier.

We landed on the small lake, idled to the south shore near the weathered old cabin, and nosed the Cessna into the bank. After unloading my gear, Scottie loaned me a radio in case of emergency.

We said our goodbyes, and Scottie left in his Cessna. As soon as the sound of his engine died away in the distance, there was nothing but the silence of the Alaskan tundra and the knowledge that soon the Arctic storms would prevail.

My plan was to make this my outpost for a week just as I had occasionally done in years past. Usually I camp on these remote streams, although on this trip, due to the lateness in the season and the potential for bad weather, I elected to stay in the little cabin. It was late October, almost November—much later than I had ever been here before.

I quickly dumped my gear on the cabin floor, slid into my waders, rigged a 6-weight fly rod, holstered my gun, and headed west, down the narrow spit of land that divides the two lakes.

I crossed the river at the outlet of the small lake and noticed it was unusually deep this year, nearly over the top of my waders. I climbed the tundra-clad bank on the far side and headed north. I thought about how alone I was, and so far from roads or trails. About 350 miles to Anchorage, and only by air. Well beyond the vestiges of civilization.

I hiked another third of a mile up the west side of the small lake to where the river emptied in. It was the place that fueled my soul. Like the crown jewel of all trout streams, it flowed out of the Alaska Range and meandered across the vast tundra, free and clean. An easy stream to wade. The bottom was gravel, and it was one of the best places in the world to hook into rainbows over ten pounds.

The flow of the river felt strong and refreshing against my waders. The wind had died, which was unusual in this place, and I felt as if the stars in the heavens had aligned.

I rigged a 15-foot leader tapered to 4X with a small pink egg pattern to imitate the pale sockeye salmon eggs the rainbows feed on in the fall. I fastened a tiny strike indicator six feet up the leader and made a long cast far upstream to give the fly time to sink deep before it was swept close to me in the current.

Almost immediately I was into big fish, many of them in the twenty-six-to-twenty-eight-inch range: rainbows so strong that they went well into the backing on my reel on their first few runs. I had to follow each of them quite a way down the river before I could land and release them.

The afternoon was wonderment. I was alone on this Alaskan river with miles and miles of fish, yet there were so many rainbows on this one long, gliding shoal that I would fish here all afternoon without leaving it.

The hours swept by like the current in the stream. Too soon the sun set low in the west. This was the time of day when the biggest rainbows stopped feeding on the sockeye eggs and began searching for something more substantial. I switched to a medium-size, black rabbit hair streamer fly, which I rigged on a full sink fly line and a short, 7-foot leader. I fished it across the current, mending the line upstream a couple of times after each cast to allow the fly to sink, then swinging it across the current below.

I was quickly hooked up to a larger rainbow that headed north like he was on a mission. He was way into my backing on the reel, and I could see him silhouetted in the distance against the Alaskan backdrop as he did a series of cartwheel-like jumps. I waded quickly to the bank and followed at a near-run through the tundra as he continued steadily upstream. He was nearly halfway to the next bend of the river before he finally decided it might be easier to go downstream. I had to race back along the game trail by the river, reeling as fast as I could to gain line. He paused in the current midstream, and finally I was able to work him over in the calmer water. When at last the big rainbow turned on his side in the shallow gravel, I slid the tape measure on him, and he was twenty-nine inches.

The sun had sunk below the distant mountains to the west as I started the hike back to the cabin. In the twilight, I saw a fox coming up the game trail toward me through the tundra. When he stopped about fifteen feet away, I too stopped. We simply eyed each other for a moment. It was an Alaskan standoff. I decided to move two or three steps off the path to see what he would do. He trotted by me on the trail and never looked back. I guess he owned the place.

After dinner I opened my pack and assembled my fly-tying gear. I do a lot of my fly tying at night, after the first day of fishing when I've had a chance to study the stream, the current, the behavior of the fish, and the temperature of the water, and after I've made hundreds of casts with time to think. Occasionally I've tied up some totally new invention, but more often it's a variation of something that already exists. I find out quickly whether my creations work because I test them out the very next day, and always there is the hope that whatever I tie will prove more effective. I'll admit that sometimes the flies I tie don't work very well, but at least I learn what doesn't work. However, occasionally I hit the jackpot—just enough to make it worth all the effort.

I tied flies for an hour or so in the cabin until I felt drowsy and climbed in my sleeping bag. Soon I was fast asleep.

It was midnight when something awakened me. I didn't know what it was; only that something didn't seem right. I opened the creaky wooden door of the cabin and peered outside. At first I thought it was the Arctic chill that had roused me, but as I looked into the night, I knew it was something more.

The lunar light sent shimmering ripples across the small lake to the front of the cabin. I glanced above to the clear Arctic heavens, and what I saw was magnificent. Strange and eerie lights stormed across the northern sky like barrages of fire in the Alaskan night, and they streaked across the sky in waves. They appeared mysterious and misty, as if ghosts were launching cosmic rays across this remote land, participants in some kind of cosmic contest, playing games with the constellations and a billion stars. It was a rarified moment in time, powerful and ethereal, and I stood there alone in the doorway of the little cabin, feeling very inconsequential.

As the northern lights cast their eerie glow across the barren landscape, something strange happened. In the west, across the small lake, there was movement in the Arctic night. It looked like a body approaching through the darkness—one at first, and then more of them—and I was alarmed. I was about to go for my gun when the realization hit me: they were caribou.

The largest herd I'd ever seen, their massive antlers bobbing and weaving in the starry light—on and on they came, more of them and more—sweeping across the tundra like a flood.

They paused where the stream enters the small lake, but only for a moment. Then they pressed forward effortlessly into the current. Like phantoms in the night they swam, ghostly in the Arctic moonlight, and their movement caused a thousand ripples on the surface of the lake. I knew that caribou are capable of swimming wide rivers; their hooves spread out somewhat when they swim, propelling them through the water like paddles with speed and grace.

These caribou glided swiftly through the small lake, reaching the shore close by to my right. The silence was broken sharply by their hooves, which clattered noisily on the rocky shoreline—a sharp contrast to the stillness of the night. Once they all reached the supple tundra, there was silence once again. Then they shook the frigid water from their wet fur bodies—heavy spray exploding from their wooly coats and guard hairs —millions of droplets glimmering like tinsel under the northern lights. I stood in the doorway of the old cabin in a trance, hoping this moment would not end. Then all at once they began to move, drifting silently away to the east, and they vanished in the night.

Gently I closed the cabin door and lit my little backpacking stove. Soon, a cup of hot chocolate was boiling. I sat alone at the weathered table and sipped the warmth of it in silence, grateful to God for allowing me to be here and bear witness to the magnificence of this day and this place, my home beyond the edge of civilization.

Well before sunrise, I woke more than a little bleary-eyed to a cold cabin. I cooked breakfast and dressed warmly for another late-October day. I wore my chest-high waders and felt-sole wading shoes, grabbed the two collapsible water buckets and my gun, and walked outside to gather drinking and cooking water for the next few days.

The old trapper's cabin sits on a narrow spit of land between the small lake in front of the cabin and the big lake behind it to the south. The big lake is called Iliamna, which in the Indigenous language of the area means "mythical great fish," a name that I feel is worthy. It's the largest lake in Alaska and arguably the most beautiful. Surrounded by snow-capped mountains, the lake is seventy-seven miles long, up to a thousand feet deep, and approximately thirty-five miles wide.

From this remote lake, the Kvichak River flows for over a hundred miles to Bristol Bay. Every July the largest run of salmon in the world comes up this river: twelve to thirteen million on average, mostly sockeyes, but also kings, pinks, chums, and silvers. When these salmon arrive, they fan out into the feeder streams where they were born, and there they spawn to produce the next generation.

Then they die.

Not most of them; all of them.

Female salmon lay between two and five thousand eggs, which means that every year, more than twenty-four billion salmon eggs are deposited in the streams feeding into and out of Lake Iliamna. The rainbow trout follow the salmon up these rivers and creeks and gorge themselves on this rich and unlimited smorgasbord of protein. The happy trout gain prodigious weight, size, and strength in the process.

Today, just as so many years in the past, I hoped to be the beneficiary of this remarkable fishery. Hurriedly I rigged my fly rod and loaded my backpack with a sandwich, chocolate bar, extra fly rod, down jacket, and drinking water. I wore my fleece and windbreaker. I checked the ammo, holstered my .475 revolver, and set off to explore the area several miles upstream.

The meandering river was full of fish, bright silver rainbows from the big lake and Kvichak River below. I was anxious to see if there were just as many fish in the deep pools upstream as there were at the shoal feeding into the small lake. I hiked for hours across the tundra to the north, almost to where the stream braids from the mass of willows, which is four or five miles from the old cabin as the crow flies—but much farther on the meander of the stream.

Way up here, toward the headwaters of this wild stream, is the stuff that dreams are made of. Time after time, I drifted and swung the little purple streamer fly across the shallow tailouts of each pool. A huge wake would appear behind the fly and follow it to midstream. Most often, a giant swirl would signify when the big rainbow had taken the fly. I would almost always see that swirl before I ever felt the fish.

I moved purposely on the tundra from pool to pool, on and on for hours, as if in a trance. It was one of those places in the world where I could fish ten or twelve hours a day, not knowing for sure why. To me this river represented freedom: it was as if I had tasted the magic of this Alaskan wilderness and now was a part of it.

At noon I stopped on a high bank overlooking the stream and pulled the sandwich from my backpack. I noticed the blue sky had turned steel gray, and I could no longer see the snow-caked mountains all around.

I explored even farther upstream, and after a while, the sky darkened, and it began to snow—large snowflakes drifting downward softly as if they were held by parachutes. It was such a beautiful sight and not at all concerning, as the temperature had dropped only slightly, and the wind was still.

But as the day grew darker, the snow fell heavier, and already there were a couple of inches of snow on the tundra. It was getting colder, too. Yet even as the temperature dropped, the big rainbows attacked my streamer fly with reckless abandon. Now I was hooking two or three huge rainbows in every pool, and they were taking the streamer fly more aggressively.

That was when it happened: the moment I felt the razor-sharp bite of a north wind that came blowing across the tundra, a sudden chill coming with it. Soon ice was forming near the banks on each side of the river, and though at first it was imperceptible, the ice was ever-so-slowly extending out into the stream from each bank.

As the snow came harder, the visibility dropped, and I knew that I would need to leave soon. I was a long way from the old trapper's cabin, and there was no way to know how bad the storm would be. But first, I wanted a few grayling for dinner.

Unfortunately Arctic grayling never bite when you want them to. I fished the shallow ripples at the heads of the pools to no avail. I really had my heart set on a fish fry that evening and wished that I had started catching dinner earlier when the weather was better.

At last, after changing to a much smaller fly, I had several good-sized grayling. Too many for one meal, but with bad weather moving in, I might not be able to fish the next day, so I kept enough fish for several days.

My hands were nearly frozen from cleaning the fish in the icy stream. I swung my arms in circles around and around like an airplane propeller, twenty times on each arm to warm them up. I dug around in my pack and found the extra pair of dry gloves. The wind was picking up, and so was the snow. The temperature was dropping so fast it was a little concerning, and I estimated it was now well down in the teens.

The cold wind blew ever stronger as I put my down jacket over the fleece with the windbreaker over all of that, along with the wool cap and gloves. I reeled up my line, grabbed my pack, and headed south, not following the stream but cutting straight across the open tundra.

Visibility was now extremely limited, but I had a compass and was careful not to wander too far west. If I erred in this blinding snow, I wanted it to be to the east, where I would hit the stream and could correct.

It's difficult to walk on tundra. Three or four miles is like walking six or eight miles on hard ground. Every step is an effort, like walking on a giant sponge and sinking four or five inches with each step. The footing is firm on top of the tussocks, but they are never spaced in a way that you can hop from one to another for very long.

But now the tundra was frozen, and the surface was hard. But the snow was sticking to the bottom of my felt-sole wading shoes, building up on the bottom of each shoe in three-inch balls of ice, making each step more difficult and treacherous. It would be easy to turn an ankle—and it was not a good time or place for that.

Even more concerning than the terrain was the worsening visibility. I could now see only sixty or seventy feet, and I worried about veering off too far to the west. I cheated to the east, and once or twice I hit the stream and corrected.

For two hours, I trudged slowly through the storm, with very little visibility and no landmark to guide me other than the compass. By now, the snow was about six or seven inches deep, and with every step, it stuck to bottom of each boot with golf-ball-sized chunks of ice. My waders were frozen, and I was caked in ice from head to toe. There was no cover on the open tundra, no trees to block the wind, which by now was howling.

When at last I saw the stream again, I followed it to where it emptied into the small lake. By this time, it had turned bitter cold. The ice was nearly all the way across the river from each bank, with only a narrow channel flowing free in the middle. I wondered if the ice would be worse where the river flowed out of the small lake. It worried me because I needed to cross the stream there to get to the cabin.

The north wind seemed even stronger as I pushed my way along the west side of the small lake, and the snow was blanketing the top of the tussocks in the tundra. The outlet was the place I was most concerned about.

It was almost dark when I reached it, and the temperature was still dropping. The wind and snow were intensifying, and my biggest concern was realized when I saw that the river had frozen all the way across.

It was difficult enough to cross here even in good weather, but now with it frozen over, how would I get across? The ice looked thick—not yet thick enough to walk on, but enough to make wading and busting my way through it problematic and dangerous.

By now, the north wind was full gale force. I eased down off the bank. With my feet on the ice, I couldn't break through. Then I gave it all my weight and broke through. The deepest part of the stream was right there. It was chest-deep—only an inch away from spilling over the top of my chest-high waders. I stood on my toes in this hole of ice. The last thing I needed was water coming over the top of my waders. Hypothermia would set in immediately in this storm.

There was no other option: this was the only way back to the cabin. But as I tried to move across, I was unable to break the ice with my body. I wasn't getting anywhere.

It was a problem that needed solving fairly quickly. Somehow I had to find a way to break the ice and be careful not to get wet. Perhaps even worse would be to slip in the current and go under the ice, which would be an even faster death.

I pulled myself out of the stream up high on the bank of the frozen tundra and felt the full force of the raging north wind. I had to find some way to get across.

I removed my extra rod case, which had been strapped to the back of my pack. It was one of my dad's old metal hand-me-down rod cases. I was glad I had packed it that day as a spare. Maybe I could use it to break the ice.

I removed the spare fly rod from the case and tied it, along with my primary fly rod, already broken down, to the top of my backpack.

I eased back into the stream and pounded the ice with the metal rod case like a hammer. I would bust the ice and move forward a few inches at a time.

It was dark now and the wind blew colder. As I moved slowly out into the river, the current under the ice became swifter. I also noticed that the ice was becoming thicker as the temperature continued to drop. It was disconcerting to look behind me, where instead of seeing an open trail of water, I saw the ice had already closed in and frozen behind me.

I was upset with myself for not having left the headwaters sooner, when the storm approached. And I thought to myself, "What if I get iced in out here in this river? What then?"

There was no other choice but to keep going, a few inches at a time. My focus was breaking the ice and nothing more. By now it was pitch-black dark, and the north wind continued at gale force. I tried to concentrate on breaking the ice and not slipping under it. At the midpoint of the river, the icy water was no longer threatening to come in over the top of my waders. But the howling wind was bitter cold, and I could no longer feel my feet or my hands. I needed to reach the cabin soon—but here in midriver, the current was stronger and pulling at me more and more. I had to be careful not to slip and get sucked under.

My rod case was starting to resemble a crumpled mess, and I was afraid it might come apart before I reached the far shore. I was about two-thirds of the way across and still nearly waist-deep. Through the flying snow and darkness, I could see the silhouette of the south bank.

Finally I was within twenty-five feet of the south bank. The water was swifter, and the top of the ice was just above my knees. Once I got even closer to the south bank and the swift water beneath the ice was knee-level or lower, I had more downward leverage to smash the metal rod case against the ice, and I made better time. Finally it was shallow enough to break the ice with my feet.

I made it to the shore. The entire crossing had taken over an hour, even though the distance couldn't have been much more than eighty feet.

I shivered nearly uncontrollably as the north wind blew the snow sideways. My .475 revolver in the holster on my hip had been underwater the whole time, and I feared it would freeze up now that it was in the brutal wind. Somehow, with hands too numb to feel much of anything, I managed to remove it from the holster to check it out. The thought of the gun freezing up was worrisome because I still had a nighttime hike on the narrow spit of land between the two lakes. I had this notion that sooner or later, on one year or the next, I would meet up with a grizzly coming the opposite direction. With no room for either of us to yield, I didn't know exactly where that would lead, but it probably wasn't a good place.

At last, I reached the cabin. It was cold and dark inside. I tried to start my cookstove with hands that had lost all feeling and arms I could not control. I was shaking so much that I fumbled and fumbled, but at last I got it lit. The warmth came slowly and then painfully into my hands. Then I lit candles, backed up against tinfoil, to radiate heat and light.

Slowly the cookstove, together with the candles, did a pretty good job of heating the little cabin. It took awhile, but I finally quit shaking.

Dinner was late that night, and I was starved. I fried one of the Arctic grayling along with sliced potatoes and onions, boiling the freeze-dried green beans. I topped it off with cup after cup of hot chocolate.

About 1 a.m. the wind was raging out of the north and violently shaking the tiny cabin. It never let up the entire night.

The next morning, I opened the cabin door and was greeted with a foot of snow and a strong north wind. I pulled my thermometer out of my backpack. It was right at zero.

Two days later, the temperature at noon had climbed to only five degrees. I had run out of grayling and was starting on my supply of freeze-dried meals. I wasn't worried, as I had brought an ample supply. I pulled out the radio and called Scottie. I didn't reach him, but I did speak to Nancy, Tim's wife.

She said they had been worried about me and that everyone in Iliamna knew "there was a dude snowed in down at the lower river."

Nancy said they had bear hunters stranded all over southwest Alaska and they were flying nonstop trying to get everyone out.

She said that Scottie had taken the floats off his Cessna for the winter and could no longer use it, but he had a Super Cub on tundra wheels to get me out. He would call me at 9:00 a.m. the next morning and let me know if the weather would allow him to fly.

That night after dinner, the wind picked up and howled with renewed strength. It had shifted out of the east. The cabin shook, creaked, and swayed to the very corners of its foundation with each major blast of wind.

As I lay in my sleeping bag, I tried to imagine Scottie landing in that gale-force wind. I had never seen a place where a plane could land on the tundra on the east side of this river, and I worried about that.

The next morning the wind shifted and came from the southeast. It warmed up a bit, but only for a while. Then it shifted out of the northwest, howling once more, and the temperature dropped again.

At 9:00 a.m., Scottie called on the radio. He said, "Hey, man, I'm coming after you. Can you be ready and have your gear out there in one hour? With all this wind, I won't be able to help you load anything. I'll just be holding the airplane down."

I said, "Scottie, where will you land? I can't cross the river."

He said, "Just take your gear to that gravelly area up high in the hills east of the cabin. You'll find a depression up there. It's small, but I can land in it."

My pack and gear were ready. I put on a fleece jacket with a down jacket over that and topped it off with the windbreaker. When I opened the cabin door, the wind nearly ripped it off the hinges. I went east of the cabin, high in the hills, to scout for the place Scottie had described.

I found it. It was up high on a ridge, and the depression had an opening to the west. Over thousands of years, the wind had carried all the tundra and earth away, leaving a layer of coarse gravel.

The wind up there was almost enough to knock me down, and I had to lean into it aggressively to stay upright. This was the only depression on the high ground that I could find, but I couldn't possibly see how a plane could land there. It was that small.

I hiked through the snow back to the cabin, closed it up, and took my gear to the low place on the ridge. I waited and watched for the faded red Super Cub, and finally I saw it. It was coming in slowly from the east into a howling west wind.

A Super Cub stalls at 38 mph. I estimated that the wind was blowing over 50 mph. I saw absolutely no way Scottie could land in an area that small. But he came in slowly toward me about thirty feet off the ground. His airspeed must have been 55 mph or so, but it looked as if he was barely moving. I worried that if the wind should suddenly slow, the plane would stall out.

But the wind didn't stop roaring, and Scottie got the Super Cub on the ground. He had somehow managed to land the plane in an area of not much more than sixty feet of gravel. Scottie got out and held the plane while I threw my gear in and climbed quickly behind the pilot seat, straddling it.

Scottie literally jumped in, simultaneously shoving the throttle forward and releasing the brake, and the second he pulled back on the wheel, we were airborne. It was like taking off in a chopper. It was a near-vertical liftoff with the plane pointed west. Who needed a runway?

I looked below, and all the world was white. I had pushed my luck and waited far too late in the fall to venture into the outback of Alaska with a fly rod. But I would be back next year—only a little earlier in October.

There was an allure to this place, the lake, the rivers, and the mountains all around it—and the size of the trout. Despite the solitude and the storms, there was something about this part of Alaska that brought me back year after year.

We headed west and then banked around to the east toward Iliamna. By the time we crossed the upper river, the wind had shifted and was coming out of the north. Even though we were headed east, Scottie had to keep the nose of the plane headed mostly north to stay on course.

We landed on the gravel runway at Iliamna. When we got inside the warm little building with Tim and Nancy, Scottie said, "That was the longest flight I ever made from the lower river. I was almost flying sideways the whole time."

Four months later, I was at work at the Tombras Group in my Knoxville office. It was a Saturday morning, and I was working alone. My phone rang, and it was a call from my good friend John Emert, who I had fished with in Alaska for many years. The second I heard his voice, I knew something wasn't right.

"Charlie, I have bad news," John said. "Scottie crashed his plane and died."

I was shaken and found it hard to listen. The bush pilot who had flown us all over southwest Alaska was gone. I tried not to think about the crash. I wanted to remember Scottie the way I had known him.

I thought of all the stories I had heard through the years in the Native village and around the hunting lodges—stories of caribou and bear hunters that Scottie and his fellow bush pilots had rescued when the northern snows came early. And while my situation was not dire, I was no less grateful.

I fly fished in the remote areas of Alaska for twenty-seven years—unguided trips in the backcountry, sometimes with friends and just as often on my own. Moments in time on the wild tundra—the times when I had felt most alive. It was an important part of me and who I was.

One evening recently, I was at my fly-tying desk rummaging through the bottom drawer, and there, under old stacks of rabbit fur and elk hair, I found a small package, wrapped in brown kraft paper, stained and worn—and in it were photographs from years gone by. I held them to the flashlight on my mobile phone, and each of them left me spellbound. In an instant, it all came rushing back—that Alaskan river flowing free across the wild tundra from the mountains crusted white in snow. And I gazed for a long time at that wild river. It was somehow a part of me yet seemed lost somewhere in time.

And for a while I was there again—by the small lake glimmering in the moonlight—and there was movement in the Arctic night, caribou gliding ghostlike on the lake, the water rippling all around, reflections from the northern lights above. And there was silence as they swam.

Then I noticed something else. There in one of the photos, high on a desolate ridge, I saw the low place, and for a moment I could almost see something more, a pilot and his faded red Super Cub, and the sky was dark, and the wind was blowing cold.

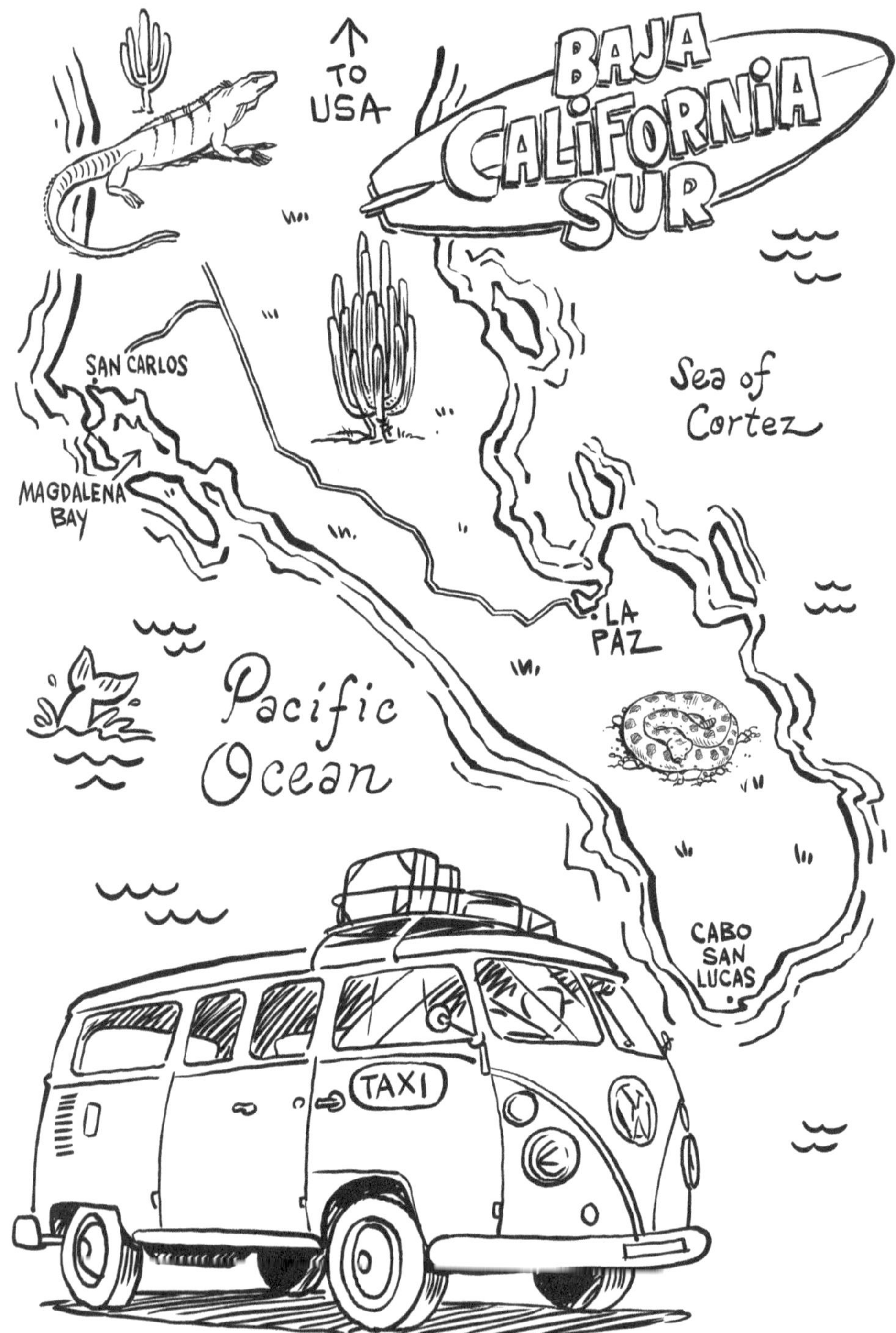
BAJA CALIFORNIA SUR
TO USA
SAN CARLOS
MAGDALENA BAY
Sea of Cortez
LA PAZ
Pacific Ocean
CABO SAN LUCAS
TAXI

NIGHT RUN ON THE BAJA, 2001

AS FAR AS the eye could see, the coastline was bleak and desolate.

The two barrier islands lay flat and windswept, offering little protection. Mostly they just hid the long bay from view.

Near the back of this lonely estuary was a small fishing village, San Carlos. Nothing much ever happened in San Carlos except for the gray whales that showed up for two weeks each winter.

I saw no sport fishing boats, just a few pangas. Magdalena Bay was seldom fished by Americans in those days, or by anyone else, because of the remote location. But the few that had fished here on excursions from Cabo or the States came back with stories of schools of striped marlin so plentiful it had stirred my imagination.

From what I had pieced together and the tales I had heard, it seemed that striped marlin congregated off the mouth of Magdalena Bay each year in a feeding frenzy and in numbers that were often equal to the Galapagos, Cocos, or New Zealand.

Reportedly, the fish were mostly in the 75 to 150-pound range. To me, these were the ideal size for a light tippet world record on a fly rod.

At the age of fifty-nine, I already held six fly rod world records on blue, white, and striped marlin, all on 16 and 20-pound test tippets. I had barely missed records on 12-pound test. While I was still near my prime, and before I quit this game, I wanted at least one ultra-light-tippet fly rod record on marlin.

I had chosen the 6-pound test category. The current record holder at that time, Tony Hedley, had landed a seventy-eight-pound striped marlin on 6-pound test tippet fishing out of Cabo. I had met Tony, who was a great fly fisherman, and his record was a magnificent catch. But I felt that with some luck, it might be possible to land one even larger, especially if the rumors of the numbers of fish in Magdalena Bay were true.

Landing a marlin on 6-pound test on a fly rod would surely involve breaking a lot of them off. That's what happens when you fly fish for marlin, even on heavier tippets. I figured that Magdalena Bay, with its huge numbers of marlin, was as good of a place as any to try for the 6-pound test fly rod record—and maybe the best.

The problem I had encountered in planning the trip was how to get a boat to this remote location on the west coast of Baja. Not just any boat would do; to have a reasonable chance of landing a marlin on 6-pound test, I needed a good boat and crew. I had heard that Jerry and Deborah Dunaway had sold their sport fishing operation and had left a friend of mine, Bubba Carter, in charge until the new owners took over. Jerry and Deborah had set so many records on conventional tackle that the Dunaway name was synonymous with big-game fishing. So was the name of the man they had left in charge.

Bubba Carter and I went back a long way together. He was from South Carolina, a native son of the Lowcountry, where his dad had been a charter boat captain. I had fly fished with Bubba on the west coast of Costa Rica and in Cocos. I remembered the time we fought a large marlin on a fly rod off the coast of Guancaste on his famous boat, *the Teireta.* We finally landed it well into the night. Needless to say, I had tremendous confidence in Bubba.

When I finally reached Bubba by phone, he was anchored off Carrillo in Costa Rica. He was excited when he learned I wanted to fish Magdalena Bay. The new owners had asked him to move the boat from Costa Rica up past Cabo and all the way to Southern California for repairs. Bubba told me he would be going right past Magdalena Bay in the fall. He said he would only have time for three days of fishing, which might reduce my number of shots at fish and chances of success.

However, it was three more days' fishing in Magdalena Bay than I was going to get any other way.

We decided it was game on for Magdalena Bay. I spent the next few months tweaking my equipment, selecting a smaller diameter fly line, backing, and shock line to cut down on drag in the water. I changed to a slightly longer fly rod with a more limber tip on a Biscayne blank. I re-snelled smaller hooks for my tube flies to make the hook set easier on the lighter leader tippet. I sent class tippets off to the IGFA to be sure they pretested legally to meet the 6-pound test limit. Everything was checked and double-checked.

Finally the November day arrived in 2001 for departure from Knoxville to Houston, from Houston to Cabo, the puddle-jumper flight from Cabo up the east coast to La Paz, and then the grueling road trip across the barren desert of Baja to Magdalena Bay on the Pacific side.

Back then, when you traveled across the center of Baja, it was hard to say for sure exactly when you would arrive. That's why the meeting time in Magdalena Bay with Bubba had been left a little vague.

The plan had been to meet Bubba sometime between dark and midnight at a little roadside restaurant. He had told me it would be the first thing I would see after I crossed the Baja desert. It was a lonely road, and the restaurant was nothing more than a taco stand with a few little folding tables and chairs outside.

It was well after dark when I arrived. Not surprisingly, there were no customers—just me, the owner, and his family. There wasn't much else in Magdalena Bay and certainly nothing as far into the darkness as I could see.

I ordered a nacho and bottled Coke and sat at one of the thin, cheap tables on a hard folding chair beside the desolate road and waited.

And waited.

Not a single car came by.

In fact, I had not seen one since I left La Paz on the east coast. It slowly occurred to me that if Bubba didn't make it to Magdalena Bay, I might be sitting at this roadside restaurant for a long time. It was not a good thing to think about.

Finally, headlights appeared in the distance, and a little Toyota truck came coughing and sputtering down the road. It pulled up near me in the dusty gravel. The tires were bare, the hood was tied down with rope, and the whole truck looked like it had been rolled once or twice a year in a cow pasture.

The driver said to me, "*Cómo estás? Eres Charlie?*" I said, "Yes."

He said, "*Muy bien. Por favor, ven conmigo.*"

I threw my duffel bag in the back of the truck and climbed into the cab with him. He said, "*Bubba me envió*," and away we went into the darkness.

Soon enough I was on the boat, sound asleep with dreams of striped marlin.

The morning of November 17, 2001, dawned glorious. Seals played in the bay near the anchor lines. The soft sun danced with the ripples on the incoming tide. Seabirds screeched overhead and dove for baitfish. Life was good, as it always is on the first day of any fishing or hunting trip.

I asked Bubba how the fishing was, but he didn't know. We were the only sport fishing boat in Magdalena Bay, and he had just arrived. We would find out soon enough.

There were three or four underwater banks in the Pacific not far off the mouth of the bay. The run by boat to the banks was short, almost shorter than the length of the bay itself, or so it seemed to me.

We came through the mouth of the bay and past one of the barrier islands. It was only 8:30 a.m. and the teaser lures were already zigzagging and diving enticingly.

I was expecting, or at least hoping for, the fast action that Magdalena Bay was known for. But we trolled one underwater bank after another for hours and hours and never raised a fish. There were a few weed lines and color changes that looked promising, but there were no fish.

I needed lots of shots if I hoped to land a striped marlin on 6-pound test tippet, but it was beginning to look like it was not going to happen. In fact, there was now the possibility that after all the preparation and planning, we might not raise any fish at all.

And then at 3:30 p.m. things changed. We suddenly raised two striped marlin at the same time. When there are two fish behind the boat, they compete, which usually lends itself to a good bite and hookup. And that's what happened.

Bubba's mate, Kevin Hibbard, pulled the teaser from the water. I cast the fly a short distance behind the boat, just beside the fish. He immediately turned, ate the fly, and for the first time in my life, I was hooked to a marlin on a fly rod with 6-pound test tippet.

I fought him for an hour or more, and he must have jumped sixty times. I was gaining confidence in my 6-pound test tippet when he suddenly accelerated on a mad dash across the surface of the Pacific toward the western sun, and the speed of his run was too much. The last time I saw him sail into the air, he was four hundred yards away and headed toward Hawaii. The speed of his run easily broke my tippet.

It was late, so we headed back in for dinner, refueling, and to develop a strategy for the next day. At dinner that evening, Bubba and I agreed that we would need at least eight or ten hookups a day to even have a chance of landing one of these strong and fast fish on 6-pound test tippet. Our plan was an early breakfast and a longer run farther into the Pacific to a different and deeper bank where, just maybe, we would find more fish.

But the second day was much like the first. We fished all morning without seeing anything. Then at 1 p.m. a marlin crashed the left long teaser again and again.

Kevin reeled the other two teasers in to get them out of the way while the other mate, Jason, teased this enraged marlin closer and closer to the boat. The sun was behind us as we looked into the depths from the transom. And there was the marlin, all lit up in a neon glow.

I threw the fly above him and to the side. He rose up from the sea and attacked it with a vengeance. In less than a second, he was off on a race across the high and sweeping Pacific waves. For the first two or three minutes, he was in the air far more than the sea.

The tail is the marlin's great motor, and this marlin's motor was in high gear. He was fast, and to my surprise, the spider-thin 6-pound tippet held during his first run and did not break.

Then he went deep, and I put light pressure on him, no more than two pounds at any time, and palmed the reel on freespool, ready for any unexpected move.

After an hour and a half of steady pull, I felt strong jolts on the line. He was deep in the ocean, shaking his head violently from side to side and trying, with his sword, to break the thin leader that barely held him. Then he rushed to the surface and began a very long run of graceful and high arching leaps. I applied the slightest possible amount of drag on the fish and bowed the rod to him on each jump to create slack with the hope that his strong and sharp dorsal fin would not cut the ultra-light leader coming across his back.

He ran and jumped and sailed through the air like a missile. He accelerated faster for four hundred yards, all this time making a long and wide circle before heading back toward me. I was reeling in slack line as fast as I could, trying to stay tight to the fish. The marlin was gaining speed, and his jumps were becoming higher and longer as he approached the boat. On the penultimate jump, he was aimed straight at me, and I was sure he was coming into the boat. I've had that happen, and I was making plans to duck behind the transom. People have been skewered and killed by such events. Fortunately for me, the two mates were alert. They saw the marlin headed to the boat and stood beside me at the transom with their gaffs. Like me, they were ready to duck behind the transom.

The marlin vaulted one last time from a high wave near the boat, and as I looked above, I thought he was surely coming down on top of me. He hung high in the air for a second, like an NBA player on the rim, and as he started downward, he veered slightly, and Kevin took a gaff shot at him. The fish exploded down into ocean inches from the transom and pulled Kevin over and downward toward the surface of the sea. The marlin would have pulled him into the ocean, but Kevin was tied into the boat, and somehow, with his feet in the air and his head down near the water, Kevin hung on to the gaff and the marlin. Jason was at his side and got his gaff into the fish, and together, they pulled this wild, violent, and still-green creature over the transom and into the boat. The marlin hit the deck with a loud crash and there was mayhem for a long time as we tried to control his savage thrashing with the two short gaffs. At last we had him under control, no one was injured, and to my amazement, the class tippet leader was still intact.

I was almost in a daze. While the fight had lasted barely over an hour and a half, it seemed to me like it was over almost as quickly as it had started. I was accustomed to long and grueling fights that last most of the day, yet here was this fish already beside me on the deck of the boat. It had happened so fast, and then it began to sink in.

The fish was ninety-four pounds. I hadn't only set the new striped marlin world record on 6-pound tippet; I had also caught the largest fish of any kind ever landed on a fly rod with a 6-pound test tippet.

I stood barefoot on the deck of the *Hooker* in a state of disbelief. My day was over. The expedition to the Baja was a success, and, remarkably, it had become so with only one shot the entire day.

Preparation and experience are crucial in fly fishing for records, but luck is sort of like a roulette wheel. Sometimes the breaks go your way and sometimes they don't. As for marlin fishing on a fly rod, I don't advise relying on luck. Ironically, I reached into my pocket and searched for the blue marlin coin. It felt good in my hand, and I thought about my dad—of the times we had spent together—and somehow in that moment I felt close to him.

It had been a dream of mine to land a large fish on such a small class tippet on a fly rod. I had been confident I could do it. I had landed billfish previously on 8-pound test tippet, but they had all been sailfish. A marlin is a much, much stronger and meaner fish, and more difficult to manage on light line. Fortunately, this marlin jumped a lot without breaking my leader and then leaped toward me at the only possible place he could be gaffed.

That night at dinner on the boat, there was much celebration and laughter. Each of us told and retold the story as we saw it from our own perspective. It was a day to remember, a fun evening, and we all turned in late.

Day three was my last day. I set my sights on the 105-pound striped marlin world record on 8-pound test tippet. Again, just as on the previous two days, I only got one shot the entire day. I landed a striped marlin that weighed in at 101 pounds, missing the current world record by only four pounds. Close, but no cigar.

By the time we weighed that fish and packed my gear, it was 9 p.m.—well after dark. I said goodbye to Bubba, Jason, and Kevin.

The Mexican in the little Toyota was there to take me back to the lonely taco stand, where I would meet a driver named Pedro willing to make a night run across the entire width of the Baja desert. I had an early morning flight from La Paz to Cabo—and back then, there was only one flight a week, so I couldn't afford to miss it.

Pedro was from La Paz and was already waiting for me. He seemed nervous. Needless to say, no one drove the road across the barren Baja at night. There was nothing on it, just desolation, rocks, dust, cacti, rattlesnakes, and darkness. With no moon, and the stars hidden by clouds, it was dark, and I had a sense of unease.

It was a six- or seven-hour trip, depending on how things went. Pedro told me he had driven it many times, but never at night. It increased my sense of apprehension when he said we would need to drive fast to make it to La Paz in time.

He drove an old 1950s-era Volkswagen van. I was amazed that it still ran. Extra gas cans were tied to the back, with water cans and two extra tires strapped down tightly inside.

We took off down the dusty, gravelly road into the pitch-blackness of the Baja. I noticed right away that Pedro's headlights didn't work on bright, and his windshield was so dirty it was hard to see the road. There were footlong divots everywhere, and he was hitting the potholes full force before he ever saw them. I thought it would be a miracle if we reached La Paz with only the two spare tires.

By midnight, we had been on the road for three hours without passing a single car or house or light of any kind. I wondered if we would go the whole distance across this desolate and lonely place without seeing anything but darkness and desert.

The Volkswagen bus squeaked and groaned and strained on the rough road, and each time we hit a pothole, it felt as if the wheels would fall off. I looked into the darkness, and as far as I could see, there was nothing. It was otherworldly, and I half expected to see a UFO.

Pedro glanced at his watch every now and then, saying we needed to hurry to reach La Paz in time. I got a glimpse at his speedometer, and it showed 80 mph. I looked up through the dirty windshield and there, broadside in front of us, not more than thirty feet away, was a massive white Brahman bull.

Pedro swerved violently to the right, and the VW bus left the road airborne at a high speed, hurtling through the Baja desert and plowing up large cacti. We must have gone two hundred feet, part of it through the air and part of it through a boulder field without touching a single rock. When the dust and smoke from the brakes settled enough to see, I looked at Pedro. He was wide-eyed, ghostly in appearance and covered in dust. He looked to me unearthly.

I must have looked equally unreal.

We inspected the bus. One tire was flat, and cacti were crammed and squeezed into every crack and cranny from impact.

Tentatively and hopefully, Pedro tried the ignition. It wouldn't start.

He tried again. It still wouldn't start.

He tried a third time, and I held my breath.

It started.

We changed the flat tire right there on the bare desert ground. Then we got the VW turned around and pointed toward the road. We searched with a flashlight to a find a path clear of rocks and boulders, and soon we were back on the dark road across the Baja desert to La Paz.

Pedro drove slower than before, and I never saw him go over 60 mph again. I'm pretty sure it was kilometers per hour and not miles per hour, so I never asked.

As we approached the outskirts of La Paz at 4:30 a.m., we passed the first vehicle we had seen all night. We pulled into the La Paz airport at 5 a.m., one hour before my 6 a.m. departure to Cabo. I gave Pedro a large tip. He had risked his life and his VW to get me to this place. Despite the mishap, I felt a bond with him, like two war buddies. We had survived it together.

As my plane lifted off and headed south down the Sea of Cortez, I watched the sunrise to my left and reflected on things.

I recalled that Saturday morning in 1978 when the IGFA world record book had arrived in the mail. What I had done on that one day had affected all my other fishing days yet to come. I had simply turned the pages of the book to see where all the biggest fish were caught.

It had been so long ago, and the journey had been circuitous—twenty years until I got my first record in Venezuela and eight more years to get this, the seventh one, here in Mexico.

I had originally thought success would come from my lifetime of fishing experience and an innate ability to fight big fish. In retrospect, it came from failures. It's how I learned and progressed. Each time I failed, I changed something in that ocean of swirling variables, always fine-tuning, adjusting, and improving. If a knot broke, the next time I tried a different one. If a connection between the fly line and the butt section of the leader gave way, the next time I rigged it differently. If the fly line caused too much drag in the water, the next time I used a thinner one. But even beyond that process of learning from failure, occasionally there was something else—something I never meant to count on: the hope of beating the odds.

In retrospect, there were many times in my life that I fished beyond my skill set, and Magdalena Bay was one of those times. No way am I good enough to consistently land a marlin on 6-pound test tippet on a fly rod with only one shot the entire day. It was just one of those days when I beat the odds, and it sure helped to have a captain like Bubba, and a mate with good reflexes like Kevin.

I thought about the night before—crossing the dark Baja and the Brahma bull in the headlights. I think maybe we beat the odds in Pedro's old Volkswagen van, too. A roll of the dice that went our way.

One thing is certain: without Pedro's lightning-quick reflexes at the wheel, I don't think I would have lived to see this fly rod record printed in the IGFA record book.

N
W
E
S
Labrador Sea
FLOWERS RIVER
GOOSE BAY
EAGLE RIVER
LABRADOR
QUEBEC

WILD ON A RIVER—FATHER AND SON, 2001

I WONDERED IF this whitewater adventure that I had dreamed of for so long would ever really happen. At the age of sixty, I had been planning and preparing for the better part of three years. Prior to that, it had only been something hoped for, but I could never quite figure out how to pull it off.

I was confident that my son Dooley and I had the ability to challenge difficult whitewater rivers that had never been run. I dreamed of the far North. In the vast Arctic, there were still a few wild rivers remaining that had never been paddled in a kayak, canoe, or raft.

For several years, I had thought maybe our adventure could be done in a kayak. However, lack of space in the kayak was the problem. The trip I envisioned required more gear than would fit inside a whitewater kayak, and I could never make the plan work.

All that changed when Dooley and I discovered whitewater canoes. Suddenly it all clicked in my mind: an expedition that would take us thousands of miles from Knoxville to the very reaches of the far North, beyond the end of the most northern road in Canada. In our whitewater canoes and with our dry bags, we could pack enough supplies and gear to make the plan work.

I had heard it said that the age of exploration on earth was over, that man had already seen everything there was to see—that there was no wilderness, desert, jungle, mountain range, or river where someone hadn't already been. Deep down I knew that this was probably true. Nevertheless, at the very least I wanted to experience a wild river that no one had run.

This whole fascination with whitewater paddling had started when I was fifty-four and Dooley was fifteen. It had begun as a father and son thing, a way for me to spend time with Dooley. I had purchased a couple of whitewater canoes, and we did a few easy backcountry fishing trips on our home rivers in the Southern Appalachians.

At first the canoes were simply a means of transportation to get us down rivers, away from people, to find more secluded camping and fishing. We escaped the problems of business and society, caught a few fish, and simply enjoyed camping in seclusion, surrounded by the beauty of our southern mountains.

Somewhere along the way, we became pretty good whitewater paddlers. After that, we began to skip the fishing and camping part of our paddling trips. Then it morphed into something more—crazy-fast rides in our open-boat canoes. Naturally we started on mild, straightforward rivers like the Hiwassee and Nantahala. Then we moved on to adrenaline-pumping class III, IV, and V—wild rivers and steep creeks. There were times at the top of some of those rapids and waterfalls when I was more than nervous, while Dooley showed nothing but quiet confidence.

Looking back on those early days of whitewater paddling, it seemed to me that we were on a mission to see just how difficult a river we could run—always looking for a bigger rapid or higher drop—continually seeking something more and more challenging. Maybe that's what this trek to the far north of Canada was all about.

For Dooley and me, those early days of whitewater paddling in the Smokies were not just occasional experiences. In the rainy season, we paddled several days a week after work or on weekends. It was cheap. It was close. We lived in the epicenter of whitewater paddling. It was all around us.

Soon Dooley passed me by, making first descents of Southern creeks that had never been run in whitewater canoes—yet he still found time to paddle with his dad on easier local waters like the Upper Tellico, the Ocoee, the Little River at the Sinks, the Upper Watauga, and the easier creeks of the Cumberland Plateau. I left the extreme steep creeks, like the West Prong of the Little Pigeon coming off the Chimneys, the Raven Fork, and others, to Dooley.

Unlike other sports where you play against a team, in whitewater paddling, there is no opponent—no Super Bowl or World Cup, and no one is keeping score. The opponent is the river itself—and a big accomplishment for a whitewater paddler is to achieve a first descent of a river that has never been run. It's one thing to attain a first descent where you are close to a road or a trail. It's quite another to pull off a first descent in true wilderness. It adds an extra element of preparation and risk. Nevertheless, that became our dream.

For several years I researched rivers in Northern Canada, searching for one that had never been run by raft, canoe, kayak, or other means. The river I saw in my mind was in a vast wilderness with no roads or trails—a river with continuous whitewater of class III and IV.

I narrowed the search down to the vast wilderness of Northeastern Canada, as those rivers had better whitewater and fewer polar bears. In fact, there was a point going east beyond which there were no polar bears.

There would need to be a lake or wide place in the river for a floatplane to land and drop us off with our canoes to start the trip. There would also need to be a viable landing spot downstream on a lake or deep place in the river where we could be picked up by a bush plane pilot at the end of the expedition.

Crucial in all this was the ability to scout the river in advance. I wanted to at least have a sense that the entire river was runnable, so I searched for a bush plane pilot who would agree to fly us low over the entire river so I could see what it had in store for us.

Northern Labrador was just such a place.

I made numerous phone calls to government officials and bush plane pilots. Information on the difficulty of the various rivers and their rapids was unreliable, as there were no whitewater paddlers in northern Labrador. In fact, there weren't many people at all in northern Labrador. Even if there had been whitewater paddlers there, it wouldn't have helped much, because we were only interested in information on rivers that had never been paddled.

The research and planning became a bit of a challenge. I was forced to rely on sketchy information from bush pilots who had flown over some of these wild rivers but who were not paddlers themselves. Despite these difficulties, I narrowed the search down to a general area about 350 miles from the end of the most northern road in Canada, which ends in Happy Valley-Goose Bay, Labrador.

Then, after all that planning, the unexpected happened: the Labrador government passed a law requiring the use of a native Labrador guide to accompany anyone entering the backcountry. After years of research and preparation, it looked like the expedition would never happen.

So I threw a Hail Mary. With the help of one of the bush plane pilots, we petitioned the Labrador government to grant a waiver in our case by calling this an exploratory trip not subject to the new law. Weeks passed. There were phone calls and letters and the typical government red tape, and I was convinced the trip would never happen. Maybe the bureaucrats wondered if we were true explorers?

Finally, at the last minute, the government of northern Labrador approved our expedition[10] with the loophole that it was an "exploration." We were on. Canoes were prepped. Supplies were ordered. Packing began. Lists were checked and double-checked.

On August 15, 2002, Dooley and I tied our canoes to the struts of a de Havilland Beaver floatplane parked beside a rickety and weathered dock. We were in Happy Valley-Goose Bay, Labrador, as far north as you can drive in Canada. The last thousand miles of road to Happy Valley-Goose Bay are dirt and gravel. From there on, there are no roads. I felt as if I was standing on the edge of a great beyond.

We were about to journey by floatplane even farther into this remote wilderness, with no idea what challenges might lie ahead. Prior to loading our dry bags into the airplane, we checked everything one last time. In the wilderness where we were headed, it would be unwise to fail to pack something.

We completed our checklists and enjoyed the remainder of the afternoon before the expedition began. People in Happy Valley-Goose Bay began to take interest in what we were doing. Small crowds gathered. They were curious where we were going. What river would we run? Was such a trip even possible in canoes? They asked questions. Some recommended against it, and some advised great caution. I was amazed that so many people were so curious.

The next morning, we were ready. We were especially cautious to be sure the canoes had been tied tightly to the struts of the airplane just above the floats. We checked and rechecked the knots. The one thing we didn't want to have happen was to have one of the canoes start coming loose and thrashing against the plane mid-flight.

Soon we were hundreds of miles past Happy Valley-Goose Bay and fly-

10. It's important to note that we ultimately made two expeditions to northern Labrador and ran three wilderness rivers, two of which had never been run at that time. This story takes place on two rivers: the Owl, which had never been run, and the Eagle, which had been run only once.

ing into one of the most remote areas in North America. The sight of all that wilderness below was more infinite than we ever imagined—it went on and on and on. And the farther we flew, the more we felt consumed by it. After so many years of planning and with the sight of that vast expanse, our senses were magnified.

At last the pilot pointed to the Eagle River in the distance, and soon, we came in low over the great river. Our plan was to paddle 120 miles, and for the sake of safety, we were going to scout the entire 120 miles of whitewater by air.

It was fascinating to see this remote river I had spent so much time studying on topographic maps. Here in the wild, it seemed totally different, and I noticed the continuousness of the rapids.

It's difficult to scout a river from an airplane even though you are only three hundred feet above it. You fly so fast that you gain only a general sense of what you are in for. Several times we saw big rapids and wished we could have stopped for a better look, but of course that was impossible. Nevertheless, from what Dooley and I could see from the aircraft, the river was mostly class III mixed in with class II and occasional class IV rapids. After flying upstream studying the Eagle, we came to the confluence of another river. The pilot veered the plane sharply to the right, and we followed it for a half hour or so to the source in an Arctic lake—and there we landed on water that was crystal clear.

We were entering one of the last great wilderness areas in North America. To me it was amazing that two guys from East Tennessee would ever find themselves here in this spot. We were about to be completely self-reliant, if only for a while, leaving the world of freeways and shopping centers behind and existing based on our paddling and wilderness skills.

For a long time, we stood there beside our canoes in silence, scanning the mountains and glens that seemed to stretch on and on and on. The vastness of it was overwhelming, and we knew there would be challenges we could not anticipate. We had yet to make a cast with our fly rods or take the first stroke with a paddle, and for the first time I realized that our biggest challenge might not be the whitewater. Our biggest challenge would more likely be the wilderness itself.

The weather was cold. Dooley and I put on our fleece, dry suits, and paddling gloves, and with the anticipation of adventure and the knowledge that we were entering the unknown, we peeled out into the first rapid.

Each river has a heartbeat and a character. Some rivers are treacherous. Some are sinister. Some are powerful, and some are pleasant. Whatever the personality of the river, you meet it face to face in the first rapid. Naturally I was apprehensive, but the river seemed to welcome us with continuous easy rapids. I wondered if this was simply a warm-up for what was to come. Ahead lay endless miles of virgin timber and canyons carved by time, and around each bend we wondered what lay beyond.

Soon the sky began to clear, and it warmed into our comfort range. Around midafternoon, we pulled our canoes onto a flat and dry boulder midstream and enjoyed a lunch of beef jerky and biscuits. A friendly breeze swept up the river, and the spray and mist from the rapid cooled us as we basked in the Labrador sun. In the distance, I think I heard a wolf howl, a reminder that this wilderness was for real and now we were a part of it—two paddlers from the Smokies, far beyond all roads and trails, on a river no man had run, and in that moment, I had never felt more alive.

We paddled mile after mile of continuous two-foot drops and easy rapids. Then the river poured into a large lake. Here we tied our canoes and fly fished for a while. We caught small brook trout, bright in their spawning colors. Their artistry and coloration were profound, as if they had been painted by some ancient Inuit spirit.

Dooley hooked into something much larger—a fish that broke his line. Then the same thing happened to me. We realized they were northern pike, and their teeth were far too sharp for our light trout tippets, so we moved on to avoid losing more of our flies. We paddled for hours across this idyllic lake, and as the sun waned, we finally located the outlet.

The next morning was cold and gray. We got a late start, but by 10 a.m. we were on the river. We knew that somewhere far ahead was the mighty Eagle and that this river would take us there. Hour after hour we paddled continuous whitewater through untouched wilderness, one small drop after another, and then suddenly, with no warning, we came around a bend, and there it was. The Eagle River.

It was big river—wide and swift. The current was strong, and we made good time. The hours seemed to fly by, and though we had paddled all day, we were fresh, alert, and strong—as if we were feeding off the energy of the rapids and the wilderness that beckoned us. The ride was exhilarating, and we didn't want to stop, but it was getting late in the day, and we needed to find a place to camp.

Atlantic salmon swirled in the dark pools—large boils and giant explosions in the current—and sometimes these magnificent fish even leaped clear of the water. They were good-sized salmon; some that I saw were in the twenty-five-pound range—not smaller, like the ten-pounder I had landed on a dry fly on our first descent down the Flowers River the year before.

I longed desperately to stop and fish, but I knew better. One of the first axioms of backcountry paddling is never to get caught too late on a whitewater river without a camping spot. One of the most dangerous situations in whitewater is to paddle in darkness, and it can happen to you on a river where there are few good dry spots to camp.

The willows along the riverbanks were dense, and the tundra was wet and swampy. Beyond the tundra, black spruce grew thick and nearly impenetrable, unwelcoming for a campsite. We paddled fast, and after two or three miles we spotted an open area in the trees.

Here in a small clearing, there was only one spot that was dry enough to pitch our two tents. Beyond that, every step we took on this wet, spongy tundra sent thousands of mosquitos and black flies out to greet us. We quickly donned our head nets for protection, finished setting up camp, and then walked through the spongy tundra to the river to see if we could catch dinner. Beside our campsite, the Eagle was wide with small boulders nearly everywhere. In the eddy below each of these rocks there was a fish, usually three or four, and we quickly caught a dinner of brook trout.

We were happy to discover that when we waded out into the shallows of the Eagle, the black flies and mosquitos left us. However, the minute we set foot on the wet tundra, they were back, and we needed head nets.

The next morning there was another revelation—a time right after daybreak when it was too cold for the flies. Then as it warmed up, there was a one-hour period when we had to defend ourselves with head nets. After that, as the sun began to rise and the day warmed, the flies were totally gone. We were elated.

So far, we were right on schedule. With one hundred miles of river in front of us and ten days until pickup, our plan was to make at least ten miles a day.

Anxious to get started, we broke camp that second morning when the mosquitos and black flies disappeared. We paddled several miles, and at midmorning we heard wolves off to our right and not far away. They were behind the willows and out of sight, though I longed to see them.

We paddled mile after mile of whitewater, and soon we settled into a good rhythm. Day by day we became a part of this uninhabited land, camping on gravel bars near the shore and leaving little if any sign of our existence.

The farther Dooley and I ventured, the more we realized this river was like no other we had ever run. It seemed alive, each set of rapids a pulse of thunder, and the land was wild. But there was something more. I felt it in the north wind sweeping clean and fresh along the mountains and infinite forests of northern fir and birch. With each breath I inhaled the wild, and with each stroke of my paddle, I felt the river's power, forever flowing strong—like some giant, enduring life force surging through the immensity of this vast land.

I knew full well that I might not pass this way again, and I wondered what lay ahead. I watched Dooley as he attacked the rapids and realized how fortunate we were to be alive on this river, apart from the world we had known back in Tennessee, yet together in a place nearly untouched since the dawn of time.

Each morning we fried country ham, freeze-dried scrambled eggs, and fresh-caught trout for breakfast. Then we loaded our dry bags in our little whitewater canoes, strapped it all down tight, and peeled out into the current, never fully knowing what to expect.

At first the rapids were mostly class II and III. But as we ventured farther into the unknown, the rapids increased in size, intensity, and velocity. Kneeling in my small whitewater canoe at water level, waves and rapids looked much larger and more violent than from high overhead in the de Havilland Beaver.

Suddenly we approached a narrow canyon, steep rock walls towering high on each side, and I felt the river accelerate. As we ventured swiftly into the canyon, we saw the entire torrent of the river explode into the left-hand wall where the river doglegged right and disappeared. Waves ten to fifteen feet high crashed and churned and slammed into one another, sending spay nearly to the top of the canyon. I felt the vibration from a quarter mile away and heard its thunder. We had no idea what mayhem awaited around that corner, but the sound was deafening, and the violence in front of us looked un-runnable. I was face to face with one of my fears: the possibility of being trapped in an un-runnable canyon in the wilderness.

We eddied out to survey the scene, and soon we were clinging precariously to the edge of a steep mountain. Dooley and I had just toiled for an hour up three hundred vertical feet of thick understory and loose scree. Our hope was to find a view farther into this steep, narrow canyon and the monstrous rapid ahead. We desperately needed to see if there was a safe line.

We were as high up this cliff-like mountain as we could safely climb, trying to catch a glimpse of the roaring rapid downstream. I could see only the middle and far side, and what I saw appeared un-runnable. The entire river erupted violently into that high rock wall. Somehow, I had missed this when we scouted from the air.

As the river roared into that undercut wall, there were holes and hydraulics large enough to swallow an SUV. It was obvious from where we had climbed to scout that the middle and left side of the river were class VI, a possible death trap.

What we couldn't get a glimpse of was the far right side of the river looking downstream. We needed to run this canyon if possible. Portaging in this place would be nearly out of the question. Normally when you encounter a wilderness rapid too dangerous to run, you walk your boat down the side of the river with ropes. It's called "lining the rapid." But here in this narrow and raging whitewater canyon, lining the rapid would be impossible.

If we couldn't run the rapid, it was going to cost us days or even weeks. We would have to go back upstream to a point where the gorge was less steep, carry our boats and all our gear over an extremely steep and high mountain through heavy and dense timber for miles, and then make our way back down the mountain below the canyon.

Sometimes that's not a big deal. But here in Labrador on this river, in this place, it was a big deal. The timber was too dense and thick. The black spruce and firs grew close together. The tree trunks were only a couple of feet apart, and the branches from each tree intertwined with the adjacent tree, making the forest virtually impenetrable. Carrying canoes up, across, and down that steep mountain in all that tangled mess would be problematic.

Running this rapid next to the right-hand canyon wall was our best hope. But I couldn't see it. Somehow that right-hand line had to be checked before committing to the river. We needed to be higher up the mountain.

The clifflike mountain we had climbed was full of loose rock, and we had gone about as far up that steep slope as we could safely climb with nothing solid to hold on to. We had no vantage point. One slip, one oversight here, could result in a three-hundred-foot fall and things could immediately and irrevocably become "complicated."

One choice was simply to go for it in the canoes and run the right side of the canyon blind. Going for it without scouting is dangerous. No one wants to depend on luck and hope for the best. And 350 miles beyond the last road or trail in Canada is not the best place in the world to gamble.

Dooley was to my right, clinging to the rocks. Somehow, he wormed his way a few feet farther up without falling.

"Dad," he shouted above the roar of the river, "I think I see a line down the right wall. I can't see all of it, but it looks like we could make it."

Slowly and carefully, we inched our way back down the steep canyon on the loose rocks, sending several on three-hundred-foot free falls near our canoes tied below. It was actually more difficult climbing down than it had been going up, because we couldn't see our footing below us—but at last we reached our canoes.

Like a deafening echo chamber, the river roared violently as we peeled out into the rapid. The waves were much higher than they looked from three hundred feet up on the mountainside. The current was accelerating rapidly, and Dooley led the way. The line he had spotted from high above was only a few feet wide and down the right canyon wall—not a lot of room for error. The melee in the middle of the river and the far left side was frightening, like paddling the rim of an active volcano.

Dooley was about thirty feet in front and strongly digging into the rapid with his paddle, aiming for the corner. I saw him disappear around it, where the canyon took its ninety-degree right turn. I hoped and prayed there was an eddy. There had to be.

I came to the corner in a split second, eddied right, and there was Dooley, smiling in the boiling eddy. And below us was easier water. Even a gravel bar farther down on river right.

We pulled our boats up high on the gravel bar and set up camp away from the flies and mosquitoes. It was almost dark, and trout were frying in the pan on glowing embers of spruce and fir. The light had faded in the west, and the riffles beside our camp sparkled in the Canadian moonlight. I was happy and grateful to be in this wild place.

Later that night, when the moon disappeared, I placed another stick on the fire and rested in my bedroll, staring almost trance-like at the small flames and listening to the sound of the rapid. To me, this little campfire by the wild river represented something more than just a campsite. I watched the dying embers and felt the warmth, and something else. Just a feeling—something too far back in time to fathom.

The next five days were the way I had dreamed the expedition would be. Each day we paddled a few miles of whitewater and fished the slower seams below each rapid. The shoals were nearly continuous, and we wore our whitewater helmets and jackets and rarely bothered to take them off, even when we fly fished, as the next rapid was always just ahead.

We discovered sunlit boulder gardens in the river where we caught seventeen- to twenty-one-inch brook trout on every cast. Sometimes if an area was particularly beautiful, we tied our canoes to the bank and waded the shoals. Other times we simply cast from the little whitewater canoes as we drifted through a pool. It was irrelevant how or where we fly fished. The fish were everywhere. We released hundreds and hundreds of trout and kept only a few of the very smallest brookies to cook in the evenings.

We spent more time fly fishing than we had meant to. How could we resist it? Then one day we looked at the map and realized it was twenty-five miles to the place we would meet the bush pilot. We still had two days until pickup, but we needed to average at least twelve and a half miles of whitewater each day. Paddling that distance of whitewater in a day is no big deal for Dooley and me if there are roads or trails. That way, if there is a problem, you can walk out for help. But here in the wilderness, a problem like an injury or wrapping a boat around a rock in a rapid becomes far more serious. So, we broke camp early the next morning, determined to make up for lost time in case we needed it.

As the sun began to rise over the forest, we peeled out into the current and paddled hard. The river accelerated and we noticed the rapids were becoming more powerful and continuous, and soon, it was one continuous class III rapid. Now, after days of paddling, we were in great physical shape and in tune with the power of the river. We attacked each rapid aggressively, pulling hard on each stroke and making great time. The river was increasing in speed, and the forest was flying by us on both sides. I had never achieved speed like that on a whitewater river. It was exhilarating, like a joy ride, yet always with an eye ahead for danger.

Kneeling on the saddles in our small canoes with our feet braced firmly against the foot pedals and our knees tucked firmly into the knee holes in the front brace, we felt solid and confident, and we attacked the main current, stroking hard, achieving speed, and rarely eddying out except once or twice to dump water. A canoe always maneuvers better and is more responsive when it is high in the water, and it sits higher if it is dry inside.

We were stroking hard through a mile-long rapid with extreme, high waves when the river shot us around a bend and there before us another large river rumbled into the Eagle from our left. I couldn't believe it: we had reached the pickup point, the place the bush pilot would fly in to pick us up. We had paddled twenty-five miles of solid and continuous class III rapids in a wilderness in six and a half hours, and we had arrived a day early.

We spotted an inviting flat and grassy area on river right where we pitched our tents. After a quick lunch, I paddled alone across the wide Eagle, probably about a quarter mile, to fly fish where the other sparkling clear river cascaded into it over a wide shoal.

With my canoe tied securely to the bank, I rigged my 5-weight fly rod with a sink tip line, tied on a small golden wet fly, and waded carefully out across the long, surging shoal, far from shore. It was surreal. Here we were alone, over 350 miles beyond the most northern road in Canada, and casting to fish that had never seen an angler.

Suddenly my line went tight. I was solidly into something strong. It made a fast run downstream about seventy-five yards, and there was little I could do to slow its run on my 6-pound tippet. I couldn't follow it downstream, as the swift water dropped off deeply below the shoal. Slowly and carefully, I worked my way left along the top of shoal a hundred yards to the shore where I had tied my canoe. By now, the fish had run even farther downstream and was about 150 yards into my backing. I ran along the shore to catch up and fight this fish, which I had yet to even see.

Not once did he jump, but I knew from his strength that he was not a monster brook trout: he had to be an Atlantic salmon. He sulked deep in the strong, dark current, but after an hour and a half, I was able to work him closer and closer to the shore. Finally, in the sunlit shallows, I got my first look at this spectacular fish. I had no net, so I just took my time, and eventually I slid him up onto the shallow gravel and he turned on his side. It was indeed an Atlantic salmon, and a nice one—a big hook-jawed male, thirty-seven inches long. I estimated him to be about twenty pounds. It was only my second Atlantic salmon on a fly. The feeling was euphoric, and it was such a good-sized salmon for a 5-weight trout rod.

That last night in the wilderness of Labrador, Dooley and I sat by our campfire, sipped coffee beneath the stars, and relived our experience—each rapid, the harrowing whitewater of the narrow canyon, the trout-filled boulder gardens, the smell of fir and black spruce, a bear that had watched us from across the river, the howl of the wolves out of sight but just beyond the willows.

We had challenged one of the greatest wilderness areas in North America, and at least for a while we had proved ourselves to be self-reliant. It was all so very different from the business world we had come from, yet the way that we worked together as a team was pretty much the same. Here in Labrador, we had scouted rapids together, set up safety lines for each other on the major rapids, gathered firewood together, and helped each other pitch camp and strap the gear into the canoes. It was no different than the way we worked together back home in Knoxville at the Tombras Group—the way a CEO and a president depend on each other and work together as a team to achieve results.

We had successfully paddled 120 miles of wilderness whitewater together on two rivers, one of which was a first descent. And we had done it on our own. No travel agencies. No guides. No rafts. No kayaks. Just two guys from East Tennessee with their little whitewater canoes. We had dreamed it, believed in it, fought for years to make it work, and best of all, we lived it. That's part of the magic of an expedition like this. The memories will never evaporate. They are hardwired within us.

Dooley and I never returned to northern Labrador. We meant to. We wanted to, but we did not. Life and business got in the way, as it often does. Our canoes are still there in Happy Valley-Goose Bay, waiting on a return that will never be.

I am blessed to have lived this wilderness trek, and most of all, to have experienced it with my son. I've often thought about my adventures in the great outdoors and wondered whether they were of any enduring worth to mankind. Did they in any way make the world a better place? Perhaps I'll never know. All I can be sure of is that this expedition to northern Labrador was of incalculable and lasting value to the father and son[11] who did it.

11. Since the two expeditions in northern Labrador, Dooley went on to win the North American whitewater canoe nationals in four different classes and placed fifth in the world in rodeo. He made fifty-six first descents in his whitewater canoe, including the most difficult one ever attempted in the southern United States, the Toxaway. He was featured in two movies filmed in three countries, *Canoe Movie* and *Canoe Movie 2: Uncharted Waters.*

LAGO
COCIBOLCA
NICARAGUA
GULF of
PAPAGAYO
COSTA RICA
GUANAMAR
GULF
of
NICOYA
SAN JOSÉ
QUEPOS
WHITE MAGIC

LAST CAST AT GUANAMAR, 2002

IT WAS ONE of my favorite places in the world. The white sand beach stretched for a mile or more.

Mountains with a mantle of tropical vegetation surrounded the circular bay, shielding the beach from strong winds. My view from high on the jungle-clad mountain was one of spectacular sunsets and smooth waves washing up on the pristine white sand below.

The sun set perfectly in the middle of this magical bay. And then sometimes, later, I saw the moon in the same place. It defined tranquility.

Half the mouth of the bay was protected by a reef where huge Pacific waves crashed in a crescendo and receded in a slow, constant, and dependable rhythm. It was a powerful sound yet pacifying, always there, serene and calming. To me it was the sound of never-ending freedom.

In the air there was tropical music, and if you followed it, you found the old Hotel Guanamar. Perched high on the mountainside in lush vegetation, the bar and pool were on an expansive teakwood deck where howler monkeys plied the jungle trees accompanied by colorful birds I could not identify. I felt the gentle breeze welcoming the early morning. I sipped the rich Costa Rican coffee and watched the entire enchanted scene unfold below.

Waiting at the table for my friends from Knoxville to join me for breakfast, I thought of how satisfying it was to be here in Guanacaste at the Hotel Guanamar again, and for a while, my mind drifted back seventeen years to a day in the mid-1980s—a chance meeting with another angler that had affected all the other days I would fly fish in saltwater.

The few boats that were here back then had just come in from fishing, and the bar at the Hotel Guanamar had been packed—mostly surfers as well as rich men from the States and South America, the type with coconut tree shirts, gold chain necklaces, and expensive girlfriends. I also recognized a more rugged type—American boat captains, some that were already iconic in the big-game fishing world along with their mates, hardened by the days at sea.

Lost in this large and noisy crowd, and unknown to each other, were two anglers here to fish for billfish in a way that most in the sport fishing industry had not yet seen. I was one of them.

A friend of mine, Jimmy Nix, who had lived in the States and had moved to Costa Rica, introduced me to the other. "There's another fly fisherman here in Costa Rica," he said. "He's from the Florida Keys and his name is Jim Gray. He goes by Harry. He holds the Atlantic blue marlin fly rod record. I want you to meet him." So he introduced us.

Harry had lived in Islamorada, and it turned out that we knew some of the same people. He had broken Billy Pate's world record on Atlantic blue marlin, and he was on a quest to set a Pacific blue marlin record as well. As it turned out, we were like-minded, and we hit it off right from the start.

Harry had a videocassette in his backpack with some amazing blue marlin action. He played it on the TV behind the bar, and about forty people gathered around, cheering and applauding. It was astonishing close-up video of ferocious blue marlin bites near the transom of a sport fishing boat. I could feel the adrenaline, and so could the crowd, who had likely never seen marlin attacking a fly. None of the fish on the video were landed, but it was the kind of fishing and challenge I had dreamed of.

We stayed there until 1:00 a.m. sharing experiences, talking about the places we had fished and the people we knew, but mostly about what it was like to fight blue marlin on a fly rod.

Harry said that he and his girlfriend, Wendy Don, lived south of there, in Quepos. He invited me to extend my trip and come fish with them, which I did. And we continued to fly fish for billfish together for many years until his health began to fail.

Harry lived life at full throttle. His enthusiasm was contagious. He was clever, and I rarely attempted to outwit him. He had a comeback for any wisecrack I made. And like me, he lived and breathed fly fishing. That's why he had moved to Quepos.

On our first day fishing together in Quepos, we used Harry's little nineteen-foot outboard that he and Wendy had named the *Sea Dog*. It was a big ocean for such a small boat, smaller than many of the bass boats on our lakes in Tennessee.

The sailfish bite that first day had been forty miles offshore. We headed out from Quepos, and soon we were all alone, out of sight of land, in an enormous ocean. The Pacific can be a monster when it wants to, and the waves had been high that day, but smooth and widely spaced. In the little nineteen-foot outboard, and with our light fly rods, I felt like I was on my way to a gun fight armed only with a BB gun.

I kept an eye on the clouds for any sign of weather—the ocean seemed like a sleeping giant. I wouldn't have given it a second thought in a larger sport fishing boat.

It was mind-boggling to see big sailfish devour our flies near the transom of Harry's little nineteen-foot outboard, close enough that we could have touched them with the tips of our fly rods. In such clear water, their size was magnified. It enabled me to study the way the billfish attacked the fly, and over the course of several seasons, we developed and tested strategies for good hookups depending on the behavior of the fish and whether it was a head-on bite, from the side, or going away.

Wendy usually piloted the little outboard while Harry and I took turns at the sailfish, and we had shot after shot at these finely honed predators—fish so robust that they reached a length just short of five feet in one year. They had owned the seas since before the age of man. The first ten minutes of fighting one on a fly rod was like hooking onto the back of a race car. Their acceleration was shocking.

For me, it was exactly the kind of "out there on the edge" fly fishing I had hoped to do someday. There were no guides, no "how to do it" videos. We were early to a sport while it was still pure—exploring, experimenting, learning by trial and error, and breaking new ground on the ocean with our fly rods. It was pure invention, and to me it felt like magic.

We tied new flies and tested different combinations of hooks, knots, shock tippets, leaders, fly lines, fly line backing, and fighting techniques. We failed fast and learned fast. It's one thing to fish as others had fished and learn from them. It's quite another to help lead the way. Every evening after fishing, we met in the Grand Escape restaurant where we talked for hours about what we had learned that day and wanted to learn.

We were two guys with the same dream: to catch world-record blue marlin on a fly rod. On the other hand, we were also competitors. It had crossed my mind that one day we might find ourselves working at cross purposes, that perhaps there would be a rift. After all, both of us had the same dream. But it was never an issue.

And how great it was to fly fish in Quepos with so many at-bats. This was where I began to move up a weight class in fighting big fish. I developed muscle memory: The innate ability to reduce line drag through the water or to reset the correct amount of reel drag in a millisecond. To conserve strength. To fight large fish with leverage, angle, and constant pressure. To anticipate their next move. To learn by trial and error exactly how much pressure each size tippet would take.

And now, seventeen years later, as I gaze off the teakwood deck here at the Hotel Guanamar, I think about the notion that when one thing happens, it affects every other thing that happens. In the years that had followed that chance meeting at the Hotel Guanamar, I had set seven fly rod world records on blue marlin, white marlin, striped marlin, and spearfish. Now, I sit here alone at the table on the teakwood deck of that same hotel recalling those events, and I am spellbound by the view of the tranquil bay and the vast Pacific.

I hear a greeting and laughter and look up as my fishing buddies from the Tombras Group, Brian Potter and Charlie Andrews, join me. It's the week of Super Bowl Sunday, a time of year that usually coincides with the peak of the billfish run in these fabled waters.

Soon enough we are thirty-five miles offshore on the *White Magic*, enjoying a fantastic day of fly fishing. It's one of those perfect days that goes by much too fast. Before I know it, the time is already nearly 4 p.m., and while I hate for a great fishing day or any day of fishing to end, there is still time for one last cast before we head in. In the glare of the low afternoon sun, we see a large bill slashing the left long teaser. At first, we think it might be a marlin, but as Gilber reels the teaser lure closer to the boat, we realize it is a very large sailfish.

Luckily it is my turn to cast, and there is a quick strike. I hook the fish, but in the glare, no one sees how big it is. Then I hear Danny from up on the bridge yelling, "Big fish! Big fish!" I have never hooked a sail this large, and he dives deep—well over a hundred feet into the depths below and never once did he come to the surface. It gets dark in Costa Rica about 6 p.m., and we are miles offshore, still fighting this great fish in total darkness. After four hours, he is finally close enough to the surface for me to begin the relentless and arduous process of pumping and leveraging him to the boat. It takes me another half hour of grit to do so.

At 8:30 p.m., approximately four and a half hours after hookup, I finally get him up beside the boat. It's tricky in the waves and darkness, but Pedro and Gilber both get their gaffs into the fish, and together they hoist him up and slide him over the gunnel where he crashes onto the wet deck of the *White Magic.* Danny turns on the spotlight from the bridge, which illuminates and seems to magnify the length of this great sailfish. It is nearly ten feet of solid muscle, with every drop of seawater glistening in the spotlight and illuminating his vivid natural colors. I am drained physically yet invigorated by the anticipation of weighing him.

Charlie, Brian, and I savor the moment. Danny fires up the *White Magic,* and we began the long run to the Hotel Guanamar in the dark. The warm night breeze feels friendly, and we celebrate our success. We are a long way from Tennessee, in a big dark ocean, and we have just accomplished what no one has ever done on 20-pound tippet on a fly rod.

We radio the hotel to give them a heads-up that we will be hanging a potential world-record sailfish from their large overhead entrance gate to weigh it.

When we arrive around 10 p.m., a large crowd has heard the news and has gathered for the weigh-in. My good friend Gary Carter, owner of nearby Las Ventanas del Mar and a holder of several conventional-tackle billfish records, is there. So is Enrico Capozzi, a friend and well-known light-line angler who set so many conventional records and who loaned me his certified scales in Venezuela back in 1994 to weigh my first world record.

My good friend Bubba Carter, Hall of Famer and captain of the famous *Tijereta,* is there to congratulate me. So is my longtime friend, Harry Gray. He had heard about it and had driven there to congratulate me. There are many others, insiders in the big-game angling world. It is one of those magical moments in time, like a great impromptu rendezvous—iconic symbols of the saltwater sport fishing inner circle, too many to name, are appearing one after another here in the middle of the Costa Rican night. It is improbable.

At last the fish is hoisted up onto the scales, and in that tense moment of waiting for the scale to settle down with a solid reading, we know for sure. I have just set a new fly rod Pacific sailfish world record at 130 pounds—my eighth world fly rod record. It is a record that still stands today, twenty-two years later.

I had hooked this great fish on the last cast of the final day.

Over the years, I've noticed that the last cast of any day of fishing is often a good one. I've thought about it a lot, and I am never sure why that is. Maybe I put some extra mojo on the last cast? More likely, my last cast is successful because fish just bite better later in the day as the light begins to fade. Or maybe it's because after I land a good fish, I think to myself, "Why not just end the day now and quit on a good note?"

But it's more than that—something that always makes me feel uneasy—the thing that lurks there in the back of my mind. It's simply knowing that in a life there is also a last cast. I try not to dwell on it, but the thought is always there, and I can't help but wonder if my next cast or the one after that will be it. I try not to think about it, but it's there just the same.

In my lifetime I have been blessed to have made many thousands of casts in some of the most storied waters of the world. God allowed me to do that for more than seventy years. Of all the places that I ever fly fished in saltwater, the one I loved most was a little place on the west coast of Costa Rica—the place we knew as Guanamar.

I've heard that the old hotel is closed now, boarded up and gone. It's been over twenty years since I cast a fly in those fabled waters, yet I remember the friends I fished with, and I recall with clarity the battles with big blue marlin and sailfish, some that took me well into the night. And I recall that final night thirty-five miles offshore—the strength and power of that magnificent sailfish—and I will always remember the magic in my final cast at Guanamar.

PART EIGHT

Ripples in Time

MOUNT ZIRKEL WILDERNESS AREA
NORTH FORK of the NORTH PLATTE RIVER
SHEEP MOUNTAIN
Lake John
LONE PINE CREEK
NORTH FORK RANCH
to WALDEN, CO

THE RANCH, 2004

I OPEN THE ranch house door, and the air is crisp and cold. I walk past the bunk house in the darkness, carrying my Thermos of coffee and a biscuit. The first light of day is dim, but enough to make out the vague and dark silhouettes of the alders and western willows.

Just ahead I see something towering above the trees—dark and ominous, a shape like something alive—too big to be real, like something prehistoric.

Not that long ago, I had seen a dinosaur footprint fossilized in the rocks in this valley of North Park, and I know that they once roamed this area where I now walk.

I approach the silhouette, and it becomes more distinct. Its chiseled teeth are the same size as those of a T-Rex and curved to the rear. Its upper jaw is huge and more cavernous.

I climb the side of this massive beast and slide into the cab. As I warm up the engine of the Hitachi excavator, I sip my coffee, enjoy the ham biscuit, and watch the morning light come slowly over the Never Summer Mountains. I sit for a while in silence. At the age of sixty-three, I have much to think about. Sometimes the thoughts are random, and it occurs to me that this beast I control is about the size of a T-Rex. I ponder the irony: two different beasts from different eras, both in this one spot.[12] One lived sixty-five

12. I was reflecting on James A. Michener's book, *Centennial* (Random House, 1974).

million years ago, yet the other rumbles across this ranch today, fueled from the other's remains. Oil and gas, perhaps from oil fields close by.

I had read somewhere that a T-Rex could deliver up to six tons of pressure in one bite. The beast I ride today can do that and more. Its body is massive, and at thirty-three thousand pounds, it is almost twice as heavy as a tyrannosaurus and reaches over thirty feet along the ground.

With its ponderous weight, the beast I control moves slowly, about three and a half miles per hour. The T-Rex was faster, somewhere around twelve and a half miles per hour, but still not nearly fast enough to catch the speeding Jeep in the movie *Jurassic Park.*

The monster I'm riding has only one arm, but it's far more powerful than the small arms of T-Rex, and it delivers a devasting blow. One touch of my finger and it responds. A flick of my hand and the machine delivers thousands of pounds of torque and pressure. With it, I am removing beaver dams from an irrigation ditch and cleaning out the ditch so the water can get to the pastures and grow grass. It takes a lot of hay to get the cattle through the winter in this high mountain valley.

I've been hard at it since early dawn, but it's something I love to do. Operating the track hoe gives me time to think and recall how my whole love affair with the Rockies began. It was a dream that had simmered in my soul for almost sixty years, since that early childhood fishing trip to Montana. Even before that, it was the very reason I'd gotten hooked on fly fishing in the first place, the thing I had left behind at the beginning and had now come back to. A place where rivers still flow pure and cold out of tall mountains, where shoals sparkle in the morning light and where clouds of mayflies flicker over sun-dappled riffles. A place to cherish the artistry and beauty of a fly cast and the hope of connecting with something wild. The mountains and rivers of the western Rockies will grow on you, and before you realize, they are wired into your being. It happens to many who fly fish, and it happened to me.

For two decades, I was wrapped up in the dream of catching big fish in the oceans and the great rivers of the world—until I realized there was something missing. Something true and pure, pulling at me gently from my past. Always there. Always haunting. For me, the yearning for it became so strong that I walked away from blue marlin fishing and the quest for records. There was nothing left to prove to myself. I just left it all behind one day and never looked back. It was time to return to my roots.

In 2002 I set out to recapture the magic from my youth—the intimate, small-stream fly fishing like my dad and I had done in the early days of the Great Smoky Mountains National Park. I hoped to share that with my grandsons one day.

It was yet another stage in my fly fishing journey. After several million casts, I had come full circle back to where I started so many years before, leaning into the joy of an intimate and seldom-fished stream, experiencing the coolness of the current forever flowing and filling all the emptiness within me, feeling the heartbeat of a wild trout in the moment I release it back into its home, and helping others discover the joy of this thing called fly fishing.

And that's how my dream to own a ranch was born. A ranch with trout water. My goal was to protect it. To care for it better than the public streams were managed. To nurture it and then let it become all that it could be. I believed that as a rancher, I could leave the land, the streams, the fish, and the wildlife a little better than when I arrived by fencing out the cattle, controlling the number of fish that were harvested, and enhancing the stream itself.

For me, it had never been about operating heavy equipment, herding cattle, mowing hayfields, and mending fences. Ranchers did all that and a whole lot more. I liked that type of rugged outdoor life and greatly admired the ranchers who did it full-time. But my dream was different. At the age of sixty-three, I simply envisioned a ranch as a place in the Rockies to retreat for a while each year. To recharge. To reminisce about an earlier life. To share experiences with family and others.

Embedded in me from earlier days in Montana were images of old ranch houses and rustic chinked log barns that stood in stark contrast to the snow-capped peaks and big, blue skies. I had never been able to get those scenes out of my head. All my life, I kept thinking maybe, just maybe, I could find a place that felt like the West was in the 1950s, or something close to it. It was something I still needed.

At first it was just a dream. Then one day I wrote a specific list of what the ideal ranch would be like, and the dream became a bit more real. The ranch I envisioned would:

Be off the beaten path and contiguous to a national forest
Have snow-capped mountain views year-round
Control a mile or more of both sides of a stream
Contain ample and relatively senior water rights

Be high enough in altitude to eliminate rattlesnakes
Have an authentic, chinked log barn and ranch house
Control all, or the majority, of the mineral rights
Have trees—cottonwood, aspen, fir, spruce, or pine
Support a wide variety of wildlife: bighorn sheep, antelope, elk, moose, bear, mule deer, mountain lion and badger

It turned out that a property like that was easy to imagine and harder to find. For two years, between 2002 and 2004, Carol and I traveled the west on the most interesting vacations where every day or two we visited a different ranch for sale. We met amazing ranchers. Some of these ranches had been in their families since the land grants, and they had the most fascinating stories to tell.

In the spring of 2004, Carol, Dooley, and I had taken a late-season ski trip to Steamboat Springs just before the ski slopes closed for the year. We skied a few days in bluebird weather with the slopes mostly to ourselves. While there, I remembered seeing a ranch advertised on a real estate website in a place called North Park, Colorado. I had never been to North Park, but the photos of this high mountain valley looked intriguing, and it was only an hour north of Steamboat, not far from the Wyoming state line.

I took some time off from skiing and drove there to see the ranch that was for sale. The four-thousand-acre ranch I visited that day was enticing, but it was too far out in the middle of the valley, not contiguous to a national forest, and didn't have the chinked log barns and buildings that give a ranch that western charm. So, I passed on it.

Word travels fast in small western towns, and Walden, Colorado, was no exception. The North Park Fishing Club had heard there was a guy in town looking for a ranch. They found me in the fly shop. Knowing the sale had fallen through and thinking that I might be a good candidate to join their club, they offered me a tour of their North Park streams, which I quickly accepted. I was interested in learning more about the miles of trout streams they had under lease.

At one point on the tour of this beautiful valley, we passed what appeared to be a deserted ranch. The chinked log ranch house and barn were in a beautiful setting where the snow-capped Continental Divide and the cliffs of Sheep Mountain contrasted with the rich grassland and meandering trout stream, which ran through this ranch for a mile. To me, it was like looking at the ranch I had seen in my dreams.

As we passed the gate to the ranch and drove the dirt road between the meandering river and Sheep Mountain, I learned it was the old Spaulding Ranch. I went to sleep that night in my room at the Antlers Inn with visions of that deserted Spaulding Ranch still in my head.

The next morning, I went to the Jackson County courthouse and discovered that the ranch had been left in a trust for four siblings. I also discovered that one of the sisters, LaVonn Spaulding Martin, was the trustee, and she and her husband, Tony Martin, were living in Yuma, Arizona (they later became great friends of ours). The courthouse had her phone number, so I called LaVonn and asked if she would allow me to spend a few hours out at the ranch to look around, because I might want to make her an offer to buy it. She agreed.

I drove the twelve miles west of town to the ranch. The weather was clear, and the temperature was comfortable. I parked my rental car, rigged up a 5-weight fly rod with a streamer fly, and waded into the North Fork of the North Platte River, which was close to the ranch house. I fished upstream for forty-five minutes and caught six brown trout from twelve to nineteen inches in length.

I drove back to town and called LaVonn. We agreed on a price contingent on approval from the other members of the trust. She said she would contact each one of them individually. I thanked her and gave her my contact information. I then drove to Steamboat to pick up Carol and Dooley for the trip back to Knoxville.

Two days later, LaVonn called and said they were all in agreement. If I would send her a contract, she said, they would all sign it.

At the time of the closing, I simultaneously purchased additional land from a nearby ranching operation and sold some of the Spaulding land I did not need back to the neighboring ranch operation in return. The net result was that Carol and I started small with 383 deeded acres, including two and a half miles of both sides of the North Fork of the North Platte River, a mile of Corral Creek, and a half mile of Lone Pine Creek.

Over the next seventeen years, I made eight additional property purchases and expanded the ranch to around fourteen hundred deeded acres and acquired mineral rights. In addition, I executed a lease from a neighboring ranch, which gave us additional fishing and recreational rights on approximately six hundred acres and another half mile of trout fishing.

The result today is a ranch of approximately two thousand acres (fourteen hundred deeded and approximately six hundred leased for recreational rights only). The east side of the ranch is accented by a rustic buck rail fence and a massive overhead log ranch gate. The ranch encompasses the north side of Sheep Mountain, the open, willow-lined irrigated fields, and both sides of the North Fork of the North Platte River. The upper part of the ranch extends through pine and aspen to the national forest. All told, the ranch has six miles of trout water. In addition to a small lake.

The old log ranch house was very small and over one hundred years old. I loved the authenticity of it. Instead of tearing it down and building something new, we added on a couple of rooms and remodeled the inside. It now features a great room with five mounts and a large stone fireplace that stretches to the ceiling. This is the most popular room in the ranch house, with most evenings after dinner spent in front of the fire.

Outside, a modest covered porch presents an unbroken panorama of the snow-capped Park Range (Sierra Madre), which is the Continental Divide, and the tall cliffs of Sheep Mountain. The Park Range holds snow year-round, and the porch rocking chairs are a popular place for early morning coffee and a view with the morning sun glistening off the peaks of white.

Shortly after purchasing the ranch, we hired a stream improvement company out of Bozeman, Montana, headed up by Joe Urbani. I had met Joe up in Pinedale, Wyoming, several years prior, and he had given me great advice concerning my search for a ranch. We remained friends, and through the years, Joe's company has enhanced more than three and a half miles of stream on the ranch. The result has been substantial improvement in the quality of the fishing. I was so impressed with the quality of the track hoe work they did that I purchased the track hoe, and on days like today, when work is needed, I operate it, which is a joy to me. Since early this morning, when I first climbed into the cab, the hours have seemed to fly by. And now it is already midafternoon. I can't explain why, but I lose all track of time when I operate this machine.

I see a flash of lightning up toward Lone Pine Creek a couple of miles away. I settle the bucket, climb out of the cab, and hop off the track to the ground. Not a good time to be in the track hoe. Just as my feet touch the ground, a bolt of lightning hits on the other side of the river behind me toward Sheep Mountain. The storm is getting closer.

It's a quarter mile to the ranch house, and the western sky is darkening with encroaching storm clouds. I walk the path through the willows along Corral Creek instead of the old two-track through the meadow by the bunkhouse. I walk slowly at first and then quicken my pace as I feel the cooling breeze that always comes before a storm.

A mule deer buck bolts from a thicket on my left and dashes toward the river. And just around the bend, in a nice long pool, I see one trout rise, and then another. The water temperature is probably dropping quickly toward their ideal range as the storm moves in.

Downstream from there, a beaver is swimming in the pool by the barn and carrying willow branches in his mouth. I see this a lot as storms approach.

It's getting windy now and the storm is coming fast from up by the track hoe. By the time I slip through the buck rail fence, hail is pelting me, lightly at first, and then pummeling hard. I'm close to the barn, so I scurry in and listen to the sound of the hail and rain hammering the roof. The tractor is in the barn and the seat is comfortable. The sound of the wind and storm makes me drowsy.

When I awake and open the door of the barn, a bull moose grazes by the bunkhouse. He gives me a look like he owns the ranch and is wondering if I have any business being there. I treat all the moose like they do indeed own the place, and maybe it's just as well that I didn't walk the two-track through the meadow by the bunkhouse.

I wait, and the moose finally ambles back to the meadow and out of sight.

I often contemplate the wonder this place. Has it cast a spell? What is it about this ranch that has a hold on me? A connection to an earlier time? Memories from Montana?

It is all of that, but there is something even more alluring—not one thing, but many. The call of a bull elk from up high in the aspens, the vastness of the sky, snowpack on the Continental Divide, riding horseback to high mountain lakes, a herd of bighorn sheep, cool mountain trout streams, tall cliffs that shine red in the afternoon sun, the wind in the high country, and the mountains and glens that are older than man. And even these are only part of the magic.

Thomas Friedman once wrote, "We live in a world where technology is accelerating faster than our ability to adapt."[13] The result is a huge void, where people struggle to interact well with each other. The ranch helps me cope with that. Here in North Park, things move slower. Maybe it's my imagination, but it seems to me that people in this high mountain valley spend less time on social media and more time actually talking with and helping one another. It's a more personal and relaxed way of life, one that I like and respect.

Another void in my life was caused by the death of my wife Carol; and often when I hurt the most, I sought comfort in that high mountain valley. Something about the majesty and beauty of this place seemed almost therapeutic. And while nothing ever took away the extreme emptiness I felt, the wildlife, the trout streams, and the work projects at the ranch made it seem more bearable.

Six and a half years after Carol passed away, I met a lady named Brenda who later became my wife. We visited the ranch with her son, John Trotter, and his son, John Charles. It was one of those early-season trips in May before the runoff. The river was clear, and the high mountains glistened, still ladened with heavy winter snow. But in the valley, at 8,100 feet, the grass was greening, the elk were calving in the willows, and mallards were thick in the watery wetlands. It was fun to share what I knew about fly fishing with John Charles and see him enjoy the outdoors with his dad. The ranch is about quality time with family.

Beyond that, the ranch is also about friends who share a love of the great outdoors and a respect for the west. Every year, I look forward to a fly fishing rendezvous of such lifelong friends from East Tennessee, including Edgar Faust, Joe Congleton, John Emert, and Graham Hunter—friends I have known and fished with for over fifty years. It's an annual pilgrimage and timed to coincide with the end of the runoff, as the water in the meadow streams begins to clear and recede. That's when the magic happens on the North Fork, and the river comes alive.

Occasionally this annual rendezvous coincides with good water levels for whitewater paddling in the rivers and steep creeks of Colorado and Wyoming. Dooley and I paddled these whitewater rivers together when I was younger, and I miss those times with my son. So it is extra special for me when Dooley

---

13. Thomas Friedman as quoted in a *WorkingNation* article by Matt Parke, August 2, 2017.

is here. I think of all the evenings at the ranch house at dinner and afterward around the big fireplace when the day's fishing tales and paddling adventures are told. That is the real allure of the ranch.

For me, the ranch is a place for family and close friends. And beyond that, it's a place to slow down and breathe the prairie wind, to stop for a while and listen to the sound of the killdeer and the Wilson's snipe as they fly softly at nightfall.

And sometimes when I'm a thousand miles away, the magic of the ranch is just to close my eyes and know it's there.

DOWN A DARK PATH WITH A FLY ROD, 2010

THE RIVER FLOWED swiftly through the meadow and the water was dark in the fading light. There was no sound except the ripples in the current.

I walked softly in the encroaching darkness as the last rays of evening light faded slowly over the Sierra Madre. Silently I approached one of my favorite pools on the North Fork. I stood quietly beside the river waiting for total darkness. That was when I first heard it—the rustling in the high grass behind me—and I knew he was there.

With his diminutive size and my weakening seventy-year-old eyesight, I could not see him. I didn't even try. And yet I knew where he was, and I knew exactly what he looked like.

He was a bit smaller than a house mouse. His tail was longer and black, and he had much bigger ears. He could hear predators coming from farther away. His eyes were larger, and he could see farther into the darkness. His fur was more brownish than a house mouse. It camouflaged him perfectly in the vegetation near the stream.

The little field mouse had another advantage over the house mouse—longer and stronger back legs. He could jump farther and climb faster. It helped him evade predators like the fox, coyote, badger, owl, bear, and mink.

With so many ways to die, I thought it was no wonder he stayed concealed in the tall grass waiting for nightfall, when he would roam the grasslands, often crossing the river in search of seeds, insects, and berries. His longer and stronger back legs helped him as he swam. How could he have known that swimming this river was like running a gauntlet? He was unaware of what lurked below. Beneath the surface were shadows of another world—and danger he could not fathom.

In the fading light, hook-jawed browns and rainbows emerged from hidden lies to ambush smaller fish and wild field mice. Massive fish that needed protein. Even in total darkness, these monster trout could detect the wake from a swimming mouse. It was the wake on the surface of the river that attracted them, and they followed it. Once they found the source of it, they slammed into the helpless mouse to disable it before swirling around violently to eat it whole. This high-protein diet led to prodigious growth rates and very large fish.

There was a time when I knew little of this. But that was before my dad and I, along with a friend from Knoxville, had flown in by floatplane to camp out on a remote outback stream on the west coast of Alaska. We had the entire fifty miles of river to ourselves. It was unmatched seclusion. We never saw another person, or any sign of one.

One day I noticed mice scurrying through the tundra and wondered if the trout fed on them. I had my fly-tying gear in the tent, and that night after dinner, I tied six deer-hair mouse flies for us to try out the next day.

Six was not enough.

By 10 a.m., the big rainbows had torn up every mouse fly I had tied. We returned to camp to tie more.

After that experience and another like it with Dooley in Labrador fishing for big brook trout, I began to use the mouse pattern everywhere I fished, even in the lower forty-eight states—especially at twilight and well into dark.

Now, years after those experiences in Alaska and Labrador, I stand here in the darkness by the North Fork. I will make no casts until I hear the coyotes howl or see the sun disappear behind the peaks on the Continental Divide.

The pool I will fish tonight is one of my favorites—a long shoal and deepening riffle sweeping into a bend and an undercut bank.

As I wait for darkness, it gives me time to think—to recall the time I fished this pool with a good friend from Knoxville, Edgar Faust. The memory is etched in my mind forever.

The morning sun had been low and behind us as we stood knee-deep in this same beautiful riffle. My rod had bowed fully from the surge of a massive fish. The big rainbow leaped high and broadside to us, and we saw him clearly, silver in the sunlight, droplets of water exploding from his broad body as he sailed airborne across the riffle. The fly pulled out in midair, and he was gone. I had never seen a rainbow that large or magnificent on this river or any other in Colorado. Edgar and I had stood there stunned and in awe.

But that was two years ago, and in the daylight. Now it is nearly pitch-black dark, and I am here in that very spot, and the memory of it gives me goosebumps.

I hear the rustling in the grass behind me. I know he is there and that I will never see him in the darkness. With his large eyes, he sees me first and scurries back into the willows. I hear him no more, and I wonder if I have inadvertently saved his life. Maybe because of seeing me he decided not to swim the gauntlet tonight.

It is darker now, and the time is right. I clip off part of my leader, leaving six feet of 12-pound test monofilament. I never fish leader this big and strong for trout in Colorado except occasionally here on this river at night, when I cast the mouse fly. In the darkness, the fish are never leader-shy.

I tie on the little brown field mouse fly. Below me, the run deepens as it glides forcefully into the bank near the willow tree. The night is black now, and I can't see the bank or the tree, not even a dark silhouette. Yet I know where the deepest part of the run is. I have fished this pool so many times in daylight that I know it well, and the darkness doesn't bother me.

I shoot some line to the bank above the willow tree and start the mouse fly moving immediately. I can't see it in the darkness, but I hear it gurgling as I swim it on the surface. That sound, and the weight and pull of the current on the rod and the line in my stripping hand, tells me the distance, and I can sense that the little brown mouse fly is swimming the surface through the deepest water next to the bank.

Suddenly I hear a tremendous explosion of water, like nothing I have heard before on a trout stream, even when fishing the mouse pattern. But I feel no take. No pull of any kind. I maintain my cool and do not set the hook because I know what will happen next. Then I hear a second gigantic swirl and a violent crash of water. I feel the solid weight of the fish. I pull back with the butt of the rod and set the hook into something gargantuan.

I feel the fish shake his enormous head viciously from side to side, each time sending the tip of my fly rod left and right a foot or more. Only a monster trout can do that.

I hear another detonation of water like a bull moose crashing into the river. The behemoth trout launches a run down the river, as fast and shocking as I have ever experienced on a trout stream. My 12-pound test leader breaks, and I feel my heart skip a beat. The sound of that leader breaking still reverberates in my ears, like the crack of a 7 Magnum. And just as before, when Edgar and I were here, I stand stunned and in awe. Only this time it is dark, and I am shaking, cold, and alone.

I often wonder if it was the same magnificent rainbow that I had hooked and lost when I had fished this pool with Edgar—or maybe it was a monster brown? I have a suspicion, but I do not know and never will.

It's been ten years since that dark night on the North Fork, and still to this day, it is the fish of a dream—the one I remember when I'm on a river as the light wanes, and I hear the little rustlings in the tall grass, and the fly I fish is the little brown field mouse that swims.

BRENDA, 2017

IT WAS EARLY MORNING before the chapel service began. People chatted busily, exchanging pleasantries, and I was introduced to a woman I didn't know.

When she looked at me, I felt as if I was looking into the sparkling eyes of an angel. She said her name. I said mine. And she said no more.

A week later I was in a small market near my home. That's when I heard her call my name. I turned, and she was there. We talked for a while and walked out of the market with plans for dinner together the next evening.

It had been more than six years since Carol had died, and this was the first time in all that while that I had entertained the idea of dinner with another woman. God guides our lives when we let him, and sometimes that happens in unexpected ways. In my case he sent Brenda, and that is how we met.

We dated. We attended parties and galas in Knoxville. We visited the ranch in Colorado with Brenda's son, John Trotter, and grandson, John Charles, and shared the magic of Colorado. Brenda and I became inseparable. We spent time at the farm near Rugby. We took walks and hunted quail behind my English Setters, Roamer and Jake. We sat in old rockers on the farmhouse porch and around the stone fireplace at night.

We enjoyed evenings with our friends at restaurants in Knoxville. Friendship became trust and trust became love and caring. Life was good, and then, in a way unimaginable, my world changed.

It was early June in 2018 the day that Brenda and I walked into a doctor's office for my appointment and test results. There had been a bone marrow biopsy to determine why my white blood cell counts were so high. When I saw the way the nurse avoided eye contact, I had a premonition that this might not be a good visit.

Normally nice and chatty, she said not a word and quickly ushered us into a patient room. We sat on the two cushioned chairs, staring at the walls. Finally came a knock at the door, and the doctor entered. I braced myself—a premonition that bad news might be coming.

The doctor looked down at the floor, coughed, shuffled his feet, and finally looked at Brenda and then me as he said, "I'm afraid I have some bad news." There was an awkward pause and he said, "You have leukemia."

I let that sink in for a moment and said, "Is it acute or chronic?" He said, "I'm afraid it is acute. Acute myeloid leukemia."

My heart nearly stopped. My face felt flushed. I knew what that meant, but I asked anyway, "How long do I have to live?"

He answered, "Unfortunately it's well advanced. I'd say four and a half months."

My mind was racing. I had heard of these conversations between doctors and patients, and I dreaded the thought that someday I might be the patient. Yet here I was, and unbelievably, I was having a conversation about dying in a matter of months. The whole conversation seemed unreal, because at that moment in time, at least up until this conversation, I had felt great. Totally normal.

I summoned what courage I could and asked about treatments that might extend life. He explained that the treatments involved extended stays in hospitals on chemotherapy drugs, and that I would lose hair and fingernails and feel weak and nauseous. He said about 30 percent of patients go into a temporary remission after these chemo treatments, but 70 percent do not. Of the patients that achieve temporary remission, it lasts only months, maybe nine months or a year, if lucky, before the leukemia returns.

I felt almost out of body, like I was only witnessing this scene instead of participating in it. You can't prepare for a conversation like this any more than you can prepare for a lightning strike, a landslide, or a stray bullet in a battle. But Brenda was there by my side, and I was glad she was there. I needed her.

I asked the doctor if there was any known cure. I remember his words.

He said, "There is only one: a bone marrow transplant." He went on to say, "I don't recommend it for you. The quality of life during and after the transplant can be pretty awful. It wouldn't be right for you, anyway. Too risky for someone your age. The chemo is intense and other factors lead to bad outcomes the older you are. You are over seventy-five, and I don't know of a good reputable hospital that would accept a transplant patient of your age."

I said, "Tell me how younger patients qualify for it."

He explained, "Hospitals that do transplants are looking for patients that are younger and that are in a temporary remission."

I asked him about the success rate, and he said, "The bone marrow transplant has only a 25 percent success rate—and even less the older you are."

The unwelcome news continued as he said, "The chemo treatment for the transplant is extremely intensive. Older patients like you can't tolerate that much chemo."

He continued, "The intent of the chemo is to kill the cells inside your bones. Once that occurs, donor stem cells are infused into you during the transplant with the hope that they graft and grow in your bones. If they don't graft, you'll have zero immunity and die."

"But if the donor cells do graft," he added, "we hope they will generate new cells that might have the ability to recognize your leukemia and kill it. In the meantime, with low or no immunity, you can die from colds, flu, pneumonia, or any type of infection."

As if that was not enough, he said, "There are other problems associated with a bone marrow transplant. Your body will often reject the donor cells, or to put it another way, the donor cells will recognize they are in the wrong body and attack your vital organs. It's called Graft Versus Host Disease (GVHD), and it occurs in 70 percent of bone marrow transplant patients. It can be ugly and quite often results in death, particularly the older you are."

I said, "Let me get this straight. I'm looking at only a 30 percent chance of a temporary remission from chemotherapy to possibly qualify for a transplant which only has a 25 percent chance of success in people younger than me. So at best, my odds of a cure are no more than 7 or 8 percent. And only if I can find a hospital that will accept someone my age for a transplant. Is that correct?" He verified that it was.

Brenda and I left his office in a state of shock. As we climbed into my truck, I felt weak. Inept. Out of control and at the mercy of God and the medical community. I was dazed, off balance, feeling almost removed from myself. My life had always been about hope, dreams, confidence and building for the future—but in the span of a few minutes, I had lost all that.

I had never thought my life would end this way. I had envisioned dying quickly and unexpectedly of a heart attack like my dad. Life was not following my script. Most of the outcomes in my life had been fortunate ones, and I had too often attributed those fortunate outcomes to choices I had made, actions I had taken. But this was different—beyond my control—and it shook me to my core. I wanted to fight it, but I knew the odds—and that part of me that knew the odds wanted to mentally prepare for what would be.

As Brenda and I left the physician's office, she said, "Don't worry, we're going to find a new doctor."

As we drove home, Brenda was so encouraging. "We're going to find a hospital that will accept you for the transplant," she said. "You can do this. You haven't come this far and survived a war in Vietnam and polio and Ranger training and all the other things you have fought for in your life for you to give up."

So we went to Vanderbilt, where they had just started a bone marrow transplant program. My one goal was to get into remission and try to qualify for a transplant and hope they would accept me regardless of my age.

My cancer doctor at Vanderbilt was not encouraging. He doubted that Vanderbilt would accept me for a transplant but said he would recommend me to their selection committee if I was successful in achieving temporary remission. Even then he warned me that I might not be accepted due to my age.

I was admitted to Vanderbilt in July and started on chemo. Brenda was by my side day and night in the hospital room, always responsive to how I felt and always there for me with empathy—the absolute embodiment of hopefulness and optimism.

Each morning, she walked to a restaurant many blocks away and brought me a sausage biscuit, a coffee, and a newspaper. We settled into a routine of chemo, intensive medication, and constant IVs with someone entering the room every few moments to check one thing or another.

After six weeks of chemo, there was a second bone marrow biopsy. The result was not good. The leukemia was still there, and I was not in remission. Then I was informed by my doctor that Vanderbilt had turned me down for the bone marrow transplant. They felt I was too old for it to be successful, plus I was not in complete remission.

I was crushed. Devastated. I wondered if my only hope for a transplant, and for life, had vanished.

Through it all, Brenda was upbeat and kept telling me, "We're going to beat this thing." I was tolerating the chemo reasonably well, so the doctors put me on a second regimen in a last-ditch effort to achieve remission and hope for a transplant at some other hospital, or at least to keep me alive a little longer.

Meanwhile Brenda and I saw Dr. David Aljadir at the Cancer Institute at the University of Tennessee Medical Cancer in Knoxville. He was extremely helpful and gave us hope. He referred us to a bone marrow transplant doctor at Memorial Sloan Kettering (MSK) in New York, one of the top leukemia hospitals in the world. The doctor's name at MSK was Christina Cho.

For us to see Dr. Cho, Brenda and I needed to get to New York. That was a problem. Due to two rounds of chemo, my immune system was at zero and any infection would have killed me. The doctors had warned me that flying commercial was too risky.

Then our lifelong friends Bill and Elisabeth Sansom offered us their private plane. Brenda and I flew safely to New York City. Without the Sansoms' kindness and generosity, I would have flown commercial, and with zero immunity I most likely never would have written this book.

Bill had lost his twin brother Bob to this disease fourteen years earlier. The three of us had been schoolmates at Bearden School from the fourth through the eighth grades. Bob had gone on to graduate first in his class from the Air Force Academy and later received his MBA from Oxford as a Rhodes Scholar, while Bill had been the number-one cadet at the Citadel. I don't know what their parents did to raise such fine men, but it was remarkable and an honor in my life to have been friends with both of them.

A follow-up trip to MSK on the Sansom plane involved another bone marrow biopsy. This test showed that this second round of chemo was causing some improvement with a smaller percentage of residual leukemia than the previous test. Unfortunately, I was still not in total remission, but I was closer than before.

Then there was the very first good news in this whole ordeal. In fact, it was wonderful news. Dr. Cho informed us that my improvement was enough that MSK had accepted me for the transplant despite my age. Now, for the first time, there was hope—although the odds were still low.

In November 2018, I checked into Memorial Sloan Kettering with Brenda at my side and started on the third and more intensive round of chemo to kill the cells within my skeletal system and make room for the new donor cells that would take their place.

Through the NMDP world donor database, we hoped to find a perfect-match donor. Sometimes the doctors find that perfect match, and sometimes they don't. But without a perfect match, transplants are less successful.

Out of hundreds of thousands of potential donors, Dr. Cho found several who were a perfect match for me. From those, she chose one: a twenty-five-year-old Polish woman. It amazes me that someone would be a donor like this for a person on another continent who they don't even know and most likely will never meet. I thank God for people like her.

On December 7, 2018, the day of the bone marrow transplant, Brenda and I were more than a little nervous. The doctors informed me that I would be the oldest leukemia patient ever transplanted at MSK. Normally a person might have second thoughts about such a risky medical procedure, but at least with a transplant there was a chance for a cure. It was better than the alternative. After eight months and three rounds of chemo, I was excited to finally have this opportunity. It was the fight of a lifetime, and even though the odds were extremely low, there was hope.

A week later, I started feeling the full and delayed effects of the transplant chemo. I remember waking up in the hospital one morning and feeling unusually tired. After three rounds of chemo, I had lost my energy and strength. There was also the loss of hair, fingernails, and toenails, not all at once, but certain and gradual. Fortunately, the nausea was solved with one little pill they gave me daily, and I was so very grateful for that pill.

I soon learned that the chemo was nothing compared to what would follow. The donor cells began to reject my body, which caused internal problems and an ugly rash. I looked scary with no hair and the rash from head to toe. I had also lost a lot of weight, dropping from 190 to 150 pounds. It took a month of intensive treatment on steroids and anti-rejection drugs for my body to overcome the Graft Versus Host Disease (GVHD). I was released from the hospital for Christmas, and for a short while, I became a daily outpatient in New York City.

During this brief time as an outpatient, Brenda and I got an apartment ten blocks from the hospital. Our first night there, a doctor called from the hospital. My blood work from that afternoon had shown a slight infection, which for a transplant patient can be lethal. The doctor wanted us to get started on a prescription immediately. It was 9:00 p.m., and the pharmacy closed at 10 p.m.

Although we had each visited New York City many times, Brenda and I had never lived there, and we were almost like strangers in this daunting place. It was a dark winter night. The wind was howling, and snow was flying. And we needed to get to the pharmacy in less than an hour. We were afraid for Brenda to get in a cab or Uber with a stranger at night, so she decided to walk. I will never forget Brenda walking out into that storm alone in that strange and often dangerous city. Not knowing her way, she walked in the wrong direction and had to backtrack. She must have been terribly cold and afraid, but she arrived at the correct pharmacy minutes before it closed. It took a lot of courage for a woman from Tennessee to walk the streets of New York City at night looking for a place she had never been. God had placed an angel there to help me.

Early each morning we went into the hospital for blood lab tests, and based on the results, we usually spent the rest of the day in the hospital with me on IVs to correct whatever problems the blood tests showed. During this time, it was great to have visitors and family come from near and far to see us.

Kreis Beall flew from Knoxville to stay with us on Christmas Eve and made it so special. The apartment was only sparsely furnished, so Kreis and Brenda made a quick trip to Bed Bath and Beyond for last-minute linens and household supplies. The line of Christmas shoppers was as long as a football field. Kreis picked up a Christmas tree and threw it into the cart. It really helped warm up the little apartment and brighten the spartan décor. We decorated the tree with the sterling silver cross ornaments that Kreis brought to us from Brenda's home in Knoxville. They were ornaments Brenda had collected for thirty years. We liked the tree so much, we left it up for our entire six-month stay in New York.

There were so many friends and family who visited: Brenda's son and grandson, John and John Charles Trotter; my son Dooley and his beautiful new wife, Kyri, who stayed a week to care for me while Brenda took a needed break; Breezy and Vance Wynn; my sister, Nancy, who came all the way from Houston and stayed a week; friends from Knoxville, including Cindi DeBusk, Kay Clayton, Tim Miller, Brian Potter; and Fred and Rhonda Moody from Allardt, Tennessee, near our farm in Rugby. Brenda's great-nieces from New York City, Alex Friedman and Caroline Heyer, also visited, as did their mother, Nancy Prebul.

We also appreciated the never-ending food packages with treats like Benton's Bacon, MoonPies, and Custom Foods entrees from my good friends Graham Hunter, Edgar Faust, and Joe Congleton.

Brenda and I will forever be grateful for the friends who visited, called, and sent care packages and cards. There are no words to adequately express how much this raised our spirits as we were so far from home.

In January the rejection disease, GVHD, came back with a vengeance, and I was readmitted to the hospital for another two months. These were the worst days of my entire experience with leukemia, but eventually the steroids and anti-rejection drugs did their work, and the GVHD began to diminish.

Finally, we got some great news—the news we had dreamed of for so long. My three-month bone marrow biopsy, post-transplant, showed I was 99.6 percent grafted with no measurable leukemia. Suddenly, GVHD didn't seem so bad. By May of 2019, the GVHD completely subsided, with no apparent permanent damage to my internal organs.

After nearly six months in New York City, I was fully discharged from MSK. Brenda and I couldn't wait to get home and see the beautiful springtime and green hills of Tennessee. Bill and Elisabeth sent their plane and were there to greet us along with their daughter Kathryn Eggleston, her husband, Tommy, and their son, Will.

As the plane landed in Knoxville, Brenda simply turned to me and smiled. I remember exactly what she said: "Charlie, the best is yet to come."

And what a homecoming it was. John and John Charles were there as well, and I have never enjoyed a trip home to Tennessee more than that one.

As I write this story, it has been four and a half years since the transplant. Glory to God, I am still in the game, and each day I ask for his guidance in my life.

There was a time when I thought of myself as self-reliant and resilient. After all, I had survived so much in my life already. But this disease was like nothing I had faced before. Without Brenda, family, friends, great doctors, a donor from Poland and the grace of God, I would not have made it through.

Looking back, I feel that my fight with acute leukemia was in many ways harder on Brenda than me. For one full year it changed every part of her life—household responsibilities, exercise, sleep, and time away from her son, grandson, and friends. For months on end, she slept in hospital rooms on little fold-out chairs that were so uncomfortable. In addition, she was saddled with managing my never-ending doctor appointments, reading about clinical trials, and organizing the most perplexing multitude of medications you could ever imagine—sometimes as many as forty or fifty pills in a day.

There are far too many people in this world suffering from cancer and other diseases. But the thing I learned and am so thankful for is that there are people like Brenda who step up and help. Few appreciate what the caregiver, spouse, friend, or partner goes through—the fatigue and confusion, even the same fears and apprehension as the patient. For the first time in my life, I fully realized how much it meant to have someone like that at my side. Throughout my fight with acute leukemia, it helped me so much to know that I was not alone—that Brenda and I were in it together.

In retrospect, I think that when two people live through a tough time and weather some major adversity together, their relationship becomes much stronger than before—and if they are in love, it becomes an unbreakable bond that lasts forever.

I know full well that without the sacrifices Brenda made, I would not be here. Through all the long hospital days and nights, she was there by my side. It was unselfish love.

I waited until the spring of 2019 to propose to Brenda. I wanted to be sure I had a reasonable chance of survival. We were married in December, one year after my transplant. It was a simple wedding at home with just our immediate family and officiated by our good friend, Tim Miller, with assistance from his wife, Jenny.

We began our marriage having already faced hardship. I think this helped our relationship grow stronger. And we needed that strength, because we were married only a week when the GVHD returned to attack my lungs. The doctors first suspected pneumonia, but as my condition worsened, it became clear it was GVHD. It happened three more times in 2020, even to the extent that the doctors put me on a ventilator during two of those hospital stays.

Maybe it was a blessing in disguise, as during one of those hospital stays, they found a blockage in my heart, and I now have four stints.

One of the hardest things I had to deal with through my battle with leukemia was the realization that for the first time in my life, I was not the man I used to be. I had always taken pride in being strong, confident, and independent. Being reliant on others was the antithesis of who I was or wanted to be.

Nevertheless, time and again Brenda did for me all the things that I was too weak to manage on my own. Through it all I learned that she was the one thing I could always count on. Her love knew no season; it was steadfast and unwavering. She was always there for me, and while it was physically difficult for her, she persevered day and night.

With low immunity following the transplant, the doctors made sure I quarantined for a year to avoid catching an infection. At the end of that year, the COVID-19 pandemic hit. The net result was that Brenda and I quarantined for a year longer than most of our friends.

There was an important side benefit to quarantining for so long: it gave us time to focus on each other. Our bond grew stronger. And slowly I came to realize that deep within me there was a certainty—something I could always count on. And whenever I reached for it, I discovered Brenda, the girl who was always there for me.

When asked, most people would say they have never seen an angel. I simply tell them that that I have seen one and I know one—and her name is Brenda.

God made us for each other. It was a gift. And we are enjoying life, and marriage—and being alive, in love and together.

A WALK IN HIGH COUNTRY, 2021

THE WIND OFF the Continental Divide is gentle today. It comes from above the tree line and across the boulder fields and rock scree, barely kissing the high alpine lakes and year-round snow fields.

I will turn eighty-one this year, but today I feel more like seventy. I don't know why, but my right knee is bending more easily. By the grace of God, I have survived polio, a war in Vietnam, prostate cancer, leukemia, a heart blockage requiring four heart stints and three retinal detachments. Not too long ago, it seemed unlikely that I would ever see this high country again. And certainly not on foot like the old days.

Yet here I am in the sage and lodgepole pine with patches of winter snow still clinging to the high, north-facing slopes. This morning, I saw the large bull moose again, the second time this week, and a surprise appearance of bighorn sheep silhouetted against the sky on outcropped rocks.

Why is my right knee not hurting? It feels almost natural again. Best not to dwell on it. Just be thankful and enjoy the magnificence of this wild land.

I had read somewhere that 95 percent of the world's wild places no longer exist—transformed instead into parking lots, condos, malls, and sprawling suburbs. I worry about this a lot, the thought that this untamed land might somehow slip away. To me it is irreplaceable, a finite resource, and I rely on it. It is where I come to recharge, to unwind. I depend on it, and it would hurt if I lost it. I try not to think about it and walk on.

The path I follow grows thin, barely discernible, like a narrow thread that meanders through the understory. But at last the forest opens up a bit and the path leads me into a cool and shaded aspen grove. It's a good place to walk with a fly rod even if I cast it only once or twice today. Like the old cowboy that never sells his saddle, I guess you could say I'm an old fly fisherman who will never turn loose of his fly rod. Ever.

Beyond the aspen grove the trail once again grows faint, nothing more than the trace of wild animals: elk, mule deer, moose, bear, and mountain lion, rightful inhabitants of this vast land. The trail winds its way ribbonlike through the forest. It leads me close to the place where Jeremy took a 6 x 5 elk on his bow and we packed it out to the four-wheeler. That had been the fall of 2010, but now it is the spring of 2020 and before the runoff.

It's the time of year when the grass lies matted and flat from the weight of heavy winter snow. Soon it will be bright green and upright, but for now, it is a mix of tawny and light browns, very soft and silent to walk on. As the trail winds downward through the woods, it reminds me of another trail down into North Carolina from Spence Field, a long way from here and so many years ago.

Little by little I begin to hear the headwaters of Lone Pine Creek. The miniature clear pools and boulders remind me of the pools that Lee and I fished on the headwaters of Eagle Creek when I was twelve, and later with Duane when the bear stole my pack from beneath my head while I slept. I think of the synchronous fireflies at Elkmont. The monster trout of the swinging bridge hole. Sometimes a single memory opens a floodgate.

I stop for a while in the shade of an aspen. Best not to overdo it. Perhaps I'll just sit here in the coolness of the glen and let Colorado seep into my soul.

There is no hurry; I have all day. Nothing more to do than enjoy this gift of wilderness. I sit totally motionless just as I do when hunting wild turkey back in the hills of East Tennessee,

When I sit perfectly still in the woods, I see so much more than I do while walking. I notice the rays of morning light streaming through the canopy of aspen and spruce above, and it reminds me of my first fly fishing trip with my dad at Elkmont in the Great Smokies. The cabin porch on Jakes Creek and the early morning sunlight streaming through the leaves and down past the hand-hewn table to the rickety old floor. Dad's wicker creel with wet grass and fresh trout. He would have liked this place up high on Lone Pine Creek, and maybe he would have asked me to build a fish camp for him on this spot.

With my back at rest against the aspen, I am comfortable and unmoving. I become a part of everything, connected to the aspen and spruce, yet invisible because I remain stock-still. The woods withhold secrets from those who are impatient. There are mysteries here that are only to be revealed to those who sit perfectly still and wait. I am now immersed in the forest and a part of it—no longer a stranger in these wild woods.

I'm not sure where they came from; it's as if they simply materialized. A small herd of mule deer on the edge of the glen not fifty feet away. I marvel at how silently and softly they move—like spirits in a primeval forest.

I pay particular attention to the fawns, brand new to this wilderness and not yet in full harmony with it. I let that thought sink in for a while. It takes me back nearly forty years to the day Dooley was born—the day we first locked eyes and he looked at me in wonderment, studying me and this new world of his with intelligence.

I recall all the Little League games, football at McCallie, how proud Carol and I were to be with Dooley on the field at half time at homecoming, of whitewater paddling with Dooley in our open boat canoes in the Smokies and beyond the most northern roads in Canada—and the voyage to the "lost world" of Cocos. I remember his graduation at Vanderbilt and his ascension in the advertising world. How much more could God have blessed me with?

Suddenly the smallest of the fawns comes closer, curious about the things around him. It reminds me of my new grandson, Rex. I look forward to watching Rex grow and perhaps one day if I'm lucky, Rex and I will walk this path together for a while.

I sit comfortably and motionless, and all too soon, the entire herd vanishes. I relax and breathe the high mountain air, consuming the scent of sage and aspen and the wild all around. High above are peaks reaching to nearly thirteen thousand feet, and the air is pure and clean.

It is my hope that my grandchildren and theirs too and the ones that come after them will have the opportunity to walk the trails that I walked and enjoy this beautiful place in the high country where I am today. I come to the wilderness for refreshment, for rejuvenation—to clear my mind. The wild places have always been there for me.

I hope the wilderness is something we can always count on, a stabilizing force in a turbulent and fast-changing world. Something magnificent and permanent for all the future generations, returning us closer to our beginnings—a feeling of connection to each other and the earth and everything, something larger than humanity.

Beyond that, when I walk in wild places, it seems to increase my sense of humility and temper my ego, which in turn helps me appreciate and respect other humans and be more neighborly. Things the world needs a lot more of today and perhaps something our children and future generations can be better at than we were.

In the words of Stewart Udall, "Go well, do well my children. Cherish sunsets, wild creatures, and wild places. Have a love affair with the wonder and beauty of the earth."

Perhaps with any luck, the wild places will remain.

Note on Wilderness

Every year the amount of wilderness left in the lower 48 states of America is shrinking. I read recently that we humans spend 95 percent of our time indoors, enclosed inside our offices, homes, stores, and vehicles, totally cut off from nature. Like the author Amy Gulick, I wonder if we have lost our ability to know nature, or—and this to me is the scary part—is it because there is so little of nature left in this country for us to know?

In truth, most of the true wilderness that remains in America is in one state, Alaska. A bit too far away for most of us to enjoy a weekend walk in the woods.

As for the lower 48 states, only 2.7 percent of the land is designated as wilderness and protected by the Wilderness Act of 1964.[14] Of course there are various definitions of "wilderness," but to me a true wilderness is a vast area of nature in its original condition—in other words, a vast area that has not been modified or impacted in some way by humanity. In my mind that

14. The Wilderness Act of 1964 is a federal land management statute meant to protect federal wilderness and to create a formal mechanism for designating wilderness. It was signed into law by President Lyndon B. Johnson on September 3, 1964.

excludes most of our national parks and national forests because of the roads, trails, campgrounds, tour buses, vehicles and buildings that serve the purpose of public access and comfort.

Apparently, the federal government mostly agrees with me that national parks and national forests are not wilderness areas. A quick look at the missions of the organizations that manage national parks, wilderness areas, and national forests confirms it.

A national park's mission is to provide recreational opportunities for the public while preserving natural, cultural, and historical resources. The aim is to balance public access and preservation of resources in the most pristine state possible. Because public access is part of the aim, there are by necessity things like roads, bridges, buildings, vehicles, campgrounds, and tour buses. The result is a kind of structured visitor experience.

Conversely, a designated wilderness area aims to prioritize solitude and primitive recreational experiences while protecting an untamed, undeveloped landscape where there has been little or no human impact. In so doing, the land remains in an untouched natural state with minimal human intervention.

As you can see, there is a huge difference in intent between a federally designated wilderness area and a national park. But what about national forests? The mission of a national forest compared to a federally designated wilderness area is even more striking.

A national forest aims to manage its land for multiple uses, including recreation, timber harvesting, grazing of livestock, and wildlife conservation. To accomplish this the national forest balances resource preservation with sustainable utilization for the benefit of the public. As a result, national forests have roads, vehicles, motorcycles, mountain bikes and bicycles, ATVs, snowmobiles, campgrounds, buildings, and bridges.

The story of this chapter took place in northwest Colorado on the edge of the Mount Zirkel Wilderness Area, which is a federally designated and protected wilderness encompassing 250 square miles, seventy remote backcountry mountain lakes, vast meadows of sage with immense forests of lodgepole pine transitioning to forests of subalpine fir and spruce and finally to arctic tundra where the wilderness area straddles the Continental Divide. The Mount Zirkel Wilderness Area has no campgrounds, no chainsaws, no roads, no tour buses, no automobiles, and no buildings or structures. It has no ATVs, no snowmobiles, no aircraft or flying drones, no big game carts, no bicycles or mountain bikes, or motorcycles—no wheels of any kind, in fact. The only permissible way to enter the Mount Zirkel wilderness is on foot, on horseback, or on showshoes or skis. It looks and feels like wilderness. Further adding to the sense of immensity, the 250 square miles of the Mount Zirkel Wilderness Area is essentially encompassed by the Medicine Bow-Routt National Forest's 2,900,000 acres.

ECHOES FROM THE RIVER, 2022

THE BIG RIVER flowed gracefully, winding through the ancient forest with a mind of its own. There were other rivers in my life, but this was my favorite—the one I dreamed of when I closed my eyes to sleep—and it was always there for me.

The water was cool and clean and flowed swiftly on the shoals that sparkled in the sunlight. It came out of the Smokies where the blue mist rose, rushing over rocks and boulders from before the age of man.

As a small boy, I wondered where the river came from, and as I grew, I began to explore, in the places where the rhododendron and the hemlocks grew and everything was wonderment. I was young, and nothing much was understood, but I began to learn from the river. And I climbed the thin path up into the headwaters and farther, to the places where the rivulets sprang forth from the soul of the earth high in the Great Smokies.

In my hand was a fly rod, even as a kid. And the light streaming through the trees, the summer wind sweeping up the river, the trout finning in the quietness of the darker pools—to a boy on his way to manhood, it was all very powerful and mysterious, and I longed to unravel the mystery.

And often in a sunlit riffle there was magic, and I tried to understand. But it seemed like something that could only be revealed to those who ventured on the river and stayed the course. As I grew, I became one of them, casting flies in a thousand sunlit pools and riffles. Along the way, there were others—friends who journeyed on the river with me and those that I loved. And there was magic in the times we spent together—and I hoped that it would last, and that I had found the place I loved where time stood still. But time was like the river, always gliding on, beckoning me to places that I couldn't know—where risks were never obvious. And I wondered what lay ahead. Like the mystery of a life. Things unfathomable.

Through it all, I loved the river. For me, it was the one thing that endured. Something I could count on. It was there for me, and I followed it to the great sea, the place where all the rivers merged—where the fish were big. And I fought them on a fly rod—sometimes for hours, and into the long nights. To me, a blue marlin fought wild and violent with speed and fury like a storm. At times my life was like that, too—one storm after another: like the day the cold front hit on that last patrol in Ranger training, the air assault into a battalion of NVA in Happy Valley, swinging wildly from a rope ladder behind a chopper a thousand feet in the air—the battles for life itself with cancer and polio—the loss of my wife, Carol, when I was devastated and lost—and my fight with acute leukemia when the doctors and a girl named Brenda helped rescue me.

And I wondered how God could send such storms without a warning—or was it that I failed to listen or simply missed the clues, like that late October afternoon on the remote tundra of Alaska, when the sky turned dark, and the wind blew cold?

Sometimes the only warning of a coming storm was the eerie silence that preceded it—a sixth sense that something wasn't right—like the day so very long ago in a steep jungle valley in Vietnam, the place we called the Eagle's Claw, when I listened to the moment and recognized the clue.

Sometimes the warnings were more obvious, like the thunder in a narrow canyon, deep in the heart of northern Labrador, and the roar of that giant rapid that seemed to me like a great storm.

Looking back on eighty-two years, there is one thing I know for sure: regardless of the river we follow in life, there will be storms. Some big. Some small. But what's important is not the storm; it's how we weather it—what we learn from the adversity, and whether we get back up and continue the journey on our river, a little stronger in the aftermath.

I believe that it is in those times of adversity that endurance and character are formed and what we believe in is put to the test. And we finally discover who we are and what we can be—where we are strong and where we are weak—and we learn the same about the others who journey on the river with us.

My greatest lessons from the river were things beyond my control—the fragility of life—and I came to realize how quickly that which is important can slip away.

Maybe that's the purpose of the journey and the challenges: to drive home the truth and bring us to our senses. To help us finally realize what we had and know for sure what's important, the things that matter in our lives—the love and heartbreak, family and friendships.

Perhaps on some distant and pleasant day, in a time beyond the storms, you'll find yourself on my river. I hope to see you there. If at first you can't find me, just look farther up the river, beyond the roads and trails, higher into the Great Smokies where the blue mist hovers. I'll be waiting for you there in the shadows beneath the cool hemlocks—the place where the shoal sparkles in the sunlight, where the fish are rising and the water is untouched by man.

EPILOGUE

AGAINST ALL ODDS

The Story of the Tombras Group, 1881–2022

Warren Dockter

CHARLIE TOMBRAS is a remarkable son of East Tennessee. A veteran of the army Rangers and the Vietnam War; a recipient of numerous military accommodations including the Bronze Star for valor; a world-record holder in fly fishing; an explorer at heart; and a successful businessman. Together with his son Dooley, Charlie helped author the rise of the Tombras Group from a small Knoxville marketing agency to the largest family-owned advertising agency in the world, with more than 490 employees and offices in New York, Buenos Aires, Atlanta, Charlotte, and of course, Knoxville. Through the pages of this book, readers will come to know Charlie, his sense of adventure, determination, and drive, as well as his deep sense of family and his love of East Tennessee. But while the stories he wrote primarily focus on his life and adventures in the great outdoors, they perhaps skirt one of the more significant stories of East Tennessee business history: the history of the Tombras Group.

In many ways the story of the Tombras Group began in 1881 on a tiny island in the Aegean Sea off the coast of Greece. Here Charlie's paternal grandfather, Peter Tombras, was born. For a time, Peter lived a simple life, working the family farm and orchard with his brother Gus and enjoying the long sun-filled afternoons of his youth. But as they grew into teenagers and toiled under the Greek sun, Peter and Gus began dreaming of another life far from their island home. Sharing a vision to chase their fortunes, they decided that their future lay in America. They left their small island for the Greek mainland to work for relatives in the grocery business in Nafplion, near the ancient Mycenaean Acropolis of Tiryns. Their goal was to raise enough money to go to America. They had nothing.

So they scrimped and saved and in 1898, Peter and Gus spent their entire life savings (roughly twenty-five dollars apiece) for passage to America. After the grueling journey from the Mediterranean through the Strait of Gibraltar and across the Atlantic, they finally landed in New York City, already home to 3.4 million people.

As they embarked on their new life, Peter and Gus wasted no time finding work. Necessity compelled it, and almost immediately, they found employment in grocery stores. Having grown up in one of the most peaceful, idyllic settings on earth, they missed the rural life, and they began to wander across the Eastern United States. They eventually settled in Chattanooga, working as cooks and dish washers in restaurants, with the goal of owning both a farm and their own restaurant one day. Eventually they did both, and the Mount Vernon Colonial Restaurant, which they owned, stayed in business at the bottom of Lookout Mountain for sixty-four years. To this day there is still a road in Chattanooga called Tombras Avenue, where Peter and Gus made their homes and had a farm.

Charlie Tombras would later visit their home on Sunday afternoons while he attended McCallie Military Academy in Chattanooga, and though his grandfather Peter had passed away, Gus was always happy to welcome Charlie over for a Sunday dinner and tell him of the Greek side of his family. It was on this homestead that Charlie's father, Charles P. Tombras Sr., had grown up. Though only Greek was spoken in the Tombras household, Charles taught himself perfect English and attended Old City High School in Chattanooga. He was an excellent student, graduated early from high school, and was offered a full scholarship to the University of Michigan. But Charles wanted to stay in East Tennessee, and with no support, he worked his way through the University of Tennessee in Knoxville.

It was at UT that Charles met Ellen Weaver. The meeting was by accident. As part of her sorority initiation, Ellen had to walk backward in all buildings on the UT campus, causing her to accidentally bump into Charles. That's how they met. Charles worked hard waiting tables by night in Greek-owned restaurants such as the Tennessean, the Quarterback, and Regas. By day he earned two degrees, one in law and one in accounting.

Charles and Ellen married and stayed in Knoxville, where Charles began working for the Knoxville Utilities Board (KUB) running their advertising department in downtown Knoxville. It was here at KUB that the vision of having a Tombras advertising firm began to develop. Charles realized as he worked for KUB that there were no advertising agencies in Tennessee at that time, and he dreamed of starting the first. But he was confronted with a harsh reality. He had neither capital nor the clients to fund a start-up, and so the Tombras agency remained a dream—until one day in late 1945.

After much deliberation, Charles approached the management at KUB and pitched an idea: if he started an agency, he asked, would it be possible for him to represent KUB and take all the administrative headaches around advertising off their plate? They agreed, and on January 2, 1946, the Tombras Group was born, with KUB as its sole client. KUB even allowed Charles to operate his one-man start-up agency in the KUB building for a few months by leasing space to him until he found suitable spot across the street. The Charlie Tombras that I know was four years old at the time.

By the 1950s, the Tombras Group had grown to fifteen people under the senior Tombras's leadership. The advertising industry at that time was in a period of remarkable growth, and along with that growth came several new ad agencies in Knoxville. By the time Charlie joined the business with his father in 1966, several Tombras employees had left for key positions in other firms, and the Tombras Group had only ten employees.

Charlie later told me that while he was so excited to join his father's business, he was also apprehensive. He felt like he did not really have a background in advertising. After all, his degree from the University of Tennessee was in journalism, not marketing or advertising. On top of that, he had spent two years in the Army, part of it in Vietnam, which Charlie feared may have delayed his growth into the business of advertising.

During the early years of Charlie's career, the other larger advertising firms in Knoxville controlled most of the major accounts. Since the Tombras Group was one of the smallest ad agencies, it handled smaller clients, and the revenue that these clients generated was insufficient to recruit top talent and compete with larger ad agencies or to sustain growth.

But this did not deter Charlie. His dream was to grow Tombras as well as keep it based in Knoxville, with the goal of Tombras being the largest agency in Tennessee. But it was clear that to achieve this goal, Tombras would need to invest in more talent. This created something of a crossroads.

On the one hand, Tombras needed capital to grow if it was going to remain in Knoxville. Charlie knew he could bring in outside investors to generate the capital for growth, and it was a tempting thought. On the other hand, Charlie had made a pledge to himself that he would never sell either the company or any interest in it. He wanted 100 percent ownership to stay in the family to protect the jobs of their employees.

The answer became obvious to Charlie. Rather than compete with the larger Knoxville firms for Knoxville business Tombras would go outside Knoxville and focus on acquiring clients across Tennessee. This strategy was a fast success.

Tombras took on Beecham Laboratories, a pharmaceutical company in Bristol that later sold to pharmaceutical giant Smith Kline. They also picked up a barbecue grill manufacturer and a zinc manufacturer in Greeneville, Tennessee. This additional income allowed Charlie to begin upgrading the quality of the creative product at Tombras before trying to compete with the larger advertising agencies. According to Charlie, this set the stage for his first big break.

In 1972, the Knoxville McDonald's operator, Litton Cochran, saw a presentation Charlie made to the Knoxville Chamber of Commerce. Litton owned all the Knoxville-area McDonald's restaurants and was a friend of Ray Kroc, the chairman of McDonald's. He was so impressed with Charlie's presentation that he visited the Tombras office the next morning, asking to speak with Charlie. After the two exchanged pleasantries, Litton said, "Charlie, lots of ad agencies have been calling on me, and we've always handled our advertising program ourselves. I was so impressed with your presentation at the chamber, I wondered if you might be interested in handling our account?"

Charlie was honored and amazed all at once. He had long admired how well run Litton's operation was, and to work with a national brand like McDonald's would be amazing. Charlie thought for a moment, and scarcely containing his excitement, he blurted out, "I've been dying to work with a client like you!" Charlie later recounted that after this exchange, Litton told him several reasons he might not want the McDonald's account, including all the challenges he would face. But Charlie was inspired by Litton's confidence in him and was determined to help his new friend, as well as to grow Tombras, and begin competing with the larger ad agencies in Knoxville.

Litton and Charlie became great friends, with Litton acting as something of a mentor for Charlie. For them, there was never a reason for a written contract. Their relationship was solely based on trust, and their business together initially operated on a handshake and mutual respect. As this relationship solidified and business grew with McDonald's, Charlie once again invested in talent, and Tombras became competitive with the larger Knoxville firms and others within the state. Tombras began to acquire additional major clients such as BlueCross BlueShield of Tennessee, Farm Bureau Insurance of Tennessee, and additional McDonald's operations in markets throughout the Southeast.

Despite this success, additional growth during this period was not easy. Owing to mergers and acquisitions across all business categories, the number of potential clients was shrinking. The economic landscape for small independent advertising agencies was precarious, and many of them began to fail. In fact, the number of advertising agencies in the United States during Charlie's career significantly contracted from forty thousand to just over thirteen thousand. Despite these business headwinds and Charlie's occasional adventures into the Smokies, Montana, and beautiful wilderness areas of the world, he could not escape his central drive to achieve his goals and grow Tombras to the largest ad agency in Tennessee. But the challenge remained the same: continuing the growth necessary to secure talent without taking on outside capital.

Fortunately, as Charlie wrestled with the way forward for Tombras, new technologies appeared. In 1982, Charlie was contacted by Donald Nelson, CEO of a telecommunications company in Chicago. They were launching a new wireless technology in telephones called cellular, and along with it, a new brand they would call US Cellular. Knoxville would be the launch market for this new company, and Donald wanted Tombras to be a part of the marketing campaign to bring people over to wireless phones. The advertising campaign by Tombras was extremely successful, and as US Cellular expanded their footprint throughout the country, Tombras continued as their national ad agency of record. This led to Tombras also landing another wireless telephone brand called SunCom, which was ultimately bought by AT&T. With Charlie at the helm, the strategy of reinvesting all the new capital into recruiting more talent continued, and it attracted more clients.

Throughout the 1980s, growth at Tombras was steady and consistent. This not only created security for the Tombras Group employees, but also made it easier to recruit good candidates from outside the Knoxville market. By scaling up in this way, Tombras was able to bring on clients like MasterCraft Boats and Malibu Boats.

By the early nineties, Tombras had a working relationship with Eastman Chemical Company, which ultimately led to Tombras becoming their global agency of record. By 2001, Tombras had also developed a relationship with NASCAR, which led to the Tombras Group representing five tracks: Bristol, New Hampshire, Charlotte, Kentucky, and Atlanta—all part of Speedway Motorsports, Inc. (SMI). By this time, the agency had made a remarkable transition from a local company to a strong regional force in the ad agency business.

Then Tombras began to enhance its creative reputation through work with the State of Tennessee Department of Transportation (TDOT). The commercials designed to deter drunk driving started getting noticed, and the agency also did breakthrough work featured in bathroom advertising in bars around the state. This campaign delivered results, receiving national acclaim, and it caught the eye of the National Highway Traffic Safety Administration officials in Washington, DC. They were planning their own national campaigns to discourage drunk driving and to encourage the use of seat belts, and they invited the Tombras Group to pitch for their national account. Suddenly Tombras was pitching against the best ad agencies in the United States, most of whom were based in New York. Typically national advertisers shunned ad agencies in the heartland of America in favor of ad agencies in New York, LA, or San Francisco, but Charlie never shied away from a fight. He pushed for the administration's business and came out on top. It was the largest client ever for Tombras to that point in time.

This really marked a turning point for the Tombras Group. Suddenly the small ad agency, which started as a one-person operation in a spare room at KUB, was now operating on the national level and working with a client that had a substantial national TV advertising program. It was a massive lift in credibility that helped put Tombras on the national radar. Once again Charlie reinvested in the agency, hiring more experienced media, creative, and research talent. By 2004, through a series of small attainable goals and believing in the power of good talent and people, Charlie achieved his ambition to make the Tombras Group the largest ad agency in Tennessee.

But there was also a more personal victory for Charlie in 2004. Just as

Charlie had joined his own father in business, Charlie's son Dooley graduated from Vanderbilt and joined the Tombras Group that year. This not only marked the moment where Charlie could work with his son, but also sent a message that the Tombras Group would continue as a family-owned, independent agency. Charlie and Dooley worked very closely together, and Dooley began running the NASCAR business. Their campaigns for the various SMI tracks had a direct and immediate effect on ticket sales.

Then another game-changing moment for Tombras came in 2007. Technology had once again completely disrupted the advertising industry. The iPhone had launched and so had Facebook. Soon this new social media platform would make the entire internet a marketing medium. It meant that the effects of an advertising campaign would be more measurable, with attribution almost instantly available. Charlie and Dooley saw the limitless potential of this technological shift long before many of their competitors did. The common wisdom of the day was that brand advertising agencies should focus on the one thing they did best and stay out of digital advertising. But Charlie and Dooley saw a tremendous change coming, and with it an opportunity. To them, it wasn't even a gamble. They understood that this was the future of the industry.

As it had always done at these moments, Tombras invested back into the company, this time into teams of website developers, technical directors, data analysts, social media specialists, and digital media planners. They hired videographers and editors and built a studio to produce the social media video content their clients would need. This allowed Tombras to become an initial leader in the use of social media.

The agency made a bold decision in 2010 to reposition around connecting data and creativity for business results. Tombras placed a big bet on data from both a technology and a talent standpoint. The agency helped pioneer the use of comprehensive marketing dashboards measuring in-store sales with data visualization that was ahead of the rest of the industry. This meaningful point of differentiation allowed the agency to break through and become competitive in more national business opportunities.

This breakthrough was the impetus for the acquisition of numerous national multi-unit retail clients, including Great Clips and Pilot Flying J Travel Centers, the largest operator of travel centers in North America. It also helped the Tombras Group land another national brand located in

Tennessee, MoonPie in Chattanooga, and led to other national consumer package goods clients like FritoLay's SunChips, Bush's Beans, Krusteaz pancake mix, Steak-umm, and Josh Cellars wines.

By 2015 Tombras was closer to becoming a nationally recognized name in the ad agency world. Charlie and Dooley hired numerous ad agency consultants from New York to see what could be done to attract higher-level talent and national-level clients to Knoxville. Charlie told me he remembered one consultant in particular who was very skeptical that an ad agency in Knoxville would have anything innovative to offer. But after spending time at Tombras, they were amazed at the operation Charlie and Dooley had built—and that it could be done outside of New York. The consultant advised them, "All you have to do is get the word out about Tombras."

Later that year the Tombras Group was selected National Small Ad Agency of the Year by *AdAge*, a major accolade in the advertising industry. This helped Tombras land the national Orangetheory Fitness account. They had several hundred fitness centers across America, and the national ad campaign from Tombras helped them grow to more than fifteen hundred locations. The investment that Charlie and Dooley had made in talent and growth was really paying off, and in 2018, *Fast Company* named the Tombras Group one of the World's Most Innovative Companies. But Tombras still needed to raise national awareness of their operation.

By this time the Tombras Group had grown to over two hundred employees, which filled their four small office buildings on Concord Street to capacity. A move would be necessary for continued growth. Committed to Knoxville, Charlie and Dooley began searching for a property downtown that would be near restaurants, condos, and entertainment to help attract the kind of top talent necessary for them to compete on a national and global level. The only building suitable was the old KUB building—the very same one Charlie's father had started his one-person operation out of in 1946. The building needed construction and remodeling to make it ready for the Tombras Group, so construction began in 2017. This was also the year BB&T Bank reached out and became a Tombras client. By 2018, Tombras moved into their new fifty-four-thousand-square-foot headquarters on Gay Street. The icing on the cake that year was that *Forbes* named their ad campaign for MoonPie the "marketing success story of the year."

Just as the Tombras Group was reaching this level of recognition, there was a devasting development. In 2018, Charlie was diagnosed with acute myeloid leukemia, an often-fatal form of cancer. Dooley had actively been president of Tombras for some time, and Alice Mathews, who had run the Washington, DC, office since 2003, was made CEO to help take the burden off Charlie and help drive the next chapter of growth for the agency. Charlie retained the role of chairman.

As Charlie stepped back to focus on his health, the agency steamed ahead. In 2019, Dooley recruited and hired one of the top chief creative officers in the world, Jeff Benjamin, which sent shock waves throughout the ad industry. Jeff had worked at major firms in New York and was now joining Tombras and moving to Knoxville. That same year, *AdAge* named the Tombras Group to their prestigious "A-list" for the first time. Tombras was now on the radar of national-level clients, which enabled the addition of clients like RE/MAX, PGA TOUR Superstore, Pernod Ricard, and the largest online direct-to-consumer grocery client in the Northeast, FreshDirect. Additionally, that year BB&T and SunTrust announced their merger to form Truist, a major national bank, and Tombras was able to land that account—meaning that the Tombras Group was handling advertising for the sixth-largest bank in America. These large wins meant that Tombras would open an office in New York.

As Charlie fought cancer, Dooley continued to grow Tombras with the acquisition of fashion retailer Altar'd State. By 2022 Charlie had come back from cancer and had reintegrated into the business as chairman. By this time Dooley was taking the agency to new heights, ushering in a new era of connecting data and creativity for business results. This led to Tombras being named 2024 Independent Agency of the Year by *AdAge*.

Today, with more than 490 employees and offices in New York, Buenos Aires, Atlanta, Charlotte, and Knoxville, Tombras is the largest family-owned advertising agency in the world, far exceeding Charlie's original ambition to become the largest in Tennessee. Tombras now handles national advertising programs for clients that account for more than twenty-five thousand brick-and-mortar retail locations and places nearly $900 million annually in media advertising for their clients.

In conversations I have had with Charlie, he said there were so many employees who contributed to his success—too many to name in this book. But there were three people in his life who he is most grateful to and who helped him the most. His father, for having the foresight and courage to start the Tombras Group and teach him the business; his wife Carol, who loved and supported him through some of his hardest times; and finally, his son Dooley, who led Tombras through the digital transformation and acquisition of national clients. Charlie told me that the greatest honor of his life was to work side by side with his father and then later with Dooley and watch him grow into a leader in the advertising industry and move Tombras forward from a regional to a national and even to an international ad agency. Charlie and Dooley both feel strongly that business leaders have a responsibility to leave a company and indeed a region better than they found it. This is certainly true for what Tombras achieved for the advertising industry as well as East Tennessee.

It is almost unthinkable that a small, local company that started as a one-person operation in a tiny office in Knoxville, Tennessee, would become the largest independent, family-owned advertising agency in the world. But it happened, and it happened here in East Tennessee. Its humble beginnings were forged out of East Tennessee relationships, trust, and hard work. At any moment throughout the rise of Tombras, it might have been more convenient to move the headquarters elsewhere. It might also have been tempting to have listened to advisors who saw Knoxville as too far out of the way. Fortunately, Charlie and Dooley have been just as committed to East Tennessee as they have been to their company. Perhaps it's because Knoxville is so near to the mountains and streams that they both love for fly fishing and whitewater paddling. One thing is certain: even as Tombras grew in national prominence, Charlie and Dooley have remained invested in the community and in working with East Tennesseans, not just in Knoxville but across the region. It is this approach that helped the Tombras Group grow in the first place.

So while the story of Tombras may have started on a tiny island off the coast of Greece, it continues as a story of East Tennessee. The story of a family whose drive, determination, and insistence on investing in talent became a force in reshaping a global industry. What could be more East Tennessee than that?

AFTERWORD

By Dooley Tombras

MAY 2025

It's my pleasure to share a few updates since Dad finished his final manuscript and offer a glimpse of what's next at Tombras.

It's been nearly eighty years since my grandfather made the bold move of opening the first advertising agency in the state of Tennessee. Dad and I share a powerful commitment to preserving his values and legacy of long-term vision over short-term thinking.

Decisions are made decisively based on what's in the long-term best interest of our clients, employees, and the agency—not based on spread sheets, stock price, or quarterly goals.

Part of our success is that Dad and I share a similar philosophy. In fact, for nearly two decades we've even shared an office together. Not many fathers and sons could do that. We believe in being entrepreneurial, innovative, and we have mission of connecting data and creativity for business results. This approach has allowed us to remain an independent agency that now places nearly a billion dollars a year in media combined with the scale of 500 employees, making Tombras the largest family-owned advertising agency in the world. Our model, along with the work and business results generated on behalf of our clients, led to Tombras being recognized as the 2025 AdAge Agency of the Year.

We've always believed in investing in the agency and in top talent, and that's not going to stop. One of the things we are most proud of is Tombras becoming a magnet to the best creative, media, digital, and analytics talent in the industry. This has allowed us to open offices in New York, Atlanta, and make our first ever international expansion in Buenos Aires, Argentina last year. All while we're still headquartered where we started and a place we're proud to call home, Knoxville.

One thing is certain, my goal is to ensure that Tombras continues as a family-owned ad agency where our people have the freedom to innovate and do the best work of their careers without short term pressure from PE or shareholders.

Perhaps one day in the future my sons, Rex and Roman, will read Dad's book and be inspired to write the sequel. Stay tuned. We're just getting started.

APPENDIX

FLY ROD WORLD RECORDS SET BY THE AUTHOR

TYPE OF FISH	TIPPET	WEIGHT	LOCATION	DATE SET
blue marlin (Atlantic)	20-pound tippet	208	Venezuela	May 1994
blue marlin (Atlantic)	16-pound tippet	208	Cape Verde	March 1998
sailfish (Pacific)	20-pound tippet	130	Costa Rica	February 2002
white marlin (Atlantic)	20-pound tippet	83	Brazil	December 1996
striped marlin (Pacific)	16-pound tippet	182	Cocos Island	August 1995
striped marlin (Pacific)	6-pound tippet	94	Baja	November 2001
striped marlin (Pacific)	20-pound tippet	185	Cocos Island	August 1995
spearfish (Atlantic)	16-pound tippet	61	Cape Verde	March 1998